SpringerBriefs in Applied Sciences and Technology

SpringerBriefs present concise summaries of cutting-edge research and practical applications across a wide spectrum of fields. Featuring compact volumes of 50 to 125 pages, the series covers a range of content from professional to academic.

Typical publications can be:

- A timely report of state-of-the art methods
- An introduction to or a manual for the application of mathematical or computer techniques
- A bridge between new research results, as published in journal articles
- A snapshot of a hot or emerging topic
- An in-depth case study
- A presentation of core concepts that students must understand in order to make independent contributions

SpringerBriefs are characterized by fast, global electronic dissemination, standard publishing contracts, standardized manuscript preparation and formatting guidelines, and expedited production schedules.

On the one hand, **SpringerBriefs in Applied Sciences and Technology** are devoted to the publication of fundamentals and applications within the different classical engineering disciplines as well as in interdisciplinary fields that recently emerged between these areas. On the other hand, as the boundary separating fundamental research and applied technology is more and more dissolving, this series is particularly open to trans-disciplinary topics between fundamental science and engineering.

Indexed by EI-Compendex, SCOPUS and Springerlink.

Songlin Sun · Jiaqi Zou · Zixuan Zou ·
Shaokang Wang

Editors

Experience of PYNQ

Tutorials for PYNQ-Z2

 Springer

Editors
Songlin Sun
School of Information and Communication
Engineering
Beijing University of Posts
and Telecommunications
Beijing, China

Zixuan Zou
Beijing University of Posts
and Telecommunications
Beijing, China

Jiaqi Zou
Beijing University of Posts
and Telecommunications
Beijing, China

Shaokang Wang
Beijing University of Posts
and Telecommunications
Beijing, China

ISSN 2191-530X   ISSN 2191-5318 (electronic)
SpringerBriefs in Applied Sciences and Technology
ISBN 978-981-19-9071-7   ISBN 978-981-19-9072-4 (eBook)
https://doi.org/10.1007/978-981-19-9072-4

This Springer imprint is published by the registered company Springer Nature Singapore Pte Ltd.
The registered company address is: 152 Beach Road, #21-01/04 Gateway East, Singapore 189721,
Singapore

Preface

This book introduces PYNQ, a Python-based framework from Xilinx® that makes it easier for users to build electronic systems on Xilinx platforms. The book covers the architecture of PYNQ, the design tools and methods, the software and hardware design approach, as well as rich experiment cases on communications, multimedia and deep learning. This book should serve as a useful guide for those getting starting with, or working with PYNQ and enable the learners to have a thorough understanding of the hardware/software co-design approaches in the area of the communication, multimedia and other information system components.

This book is organized based on the teaching materials of "hardware comprehensive experiments" which is an experimental course at Beijing University of Posts and Telecommunications, for both undergraduate and graduate students, as well as domestic and international students. This course has more than ten years of teaching experience and has taught hundreds of students.

We would like to offer our special thanks to Xilinx® that provided us with various materials and experiment examples. We would not be able to get the book done without the great support of Xilinx®.

Beijing, China
August 2022

Songlin Sun
Jiaqi Zou
Zixuan Zou
Shaokang Wang

Contents

PYNQ Introduction

Abstract PYNQ is an open-source project from Xilinx. It provides a Jupyter-based framework with Python APIs for using Xilinx platforms. PYNQ supports Zynq and Zynq Ultrascale+, Zynq RFSoC, Alveo and AWS-F1 instances. PYNQ enables architects, engineers and programmers who design embedded systems to use Zynq devices, without having to use ASIC-style design tools to design programmable logic circuits [1].

What's PYNQ?

PYNQ Background

Programmable logic circuits are presented as hardware libraries called overlays. These overlays are analogous to software libraries. A software engineer can select the overlay that best matches their application. The overlay can be accessed through a Python API. Creating a new overlay still requires engineers with expertise in designing programmable logic circuits. The key difference, however, is the build once, reuse many times paradigm. Overlays, like software libraries, are designed to be configurable and reused as often as possible in many different applications.

PYNQ supports Python for programming both the embedded processors and the overlays. Python is a "productivity-level" language. To date, C and C++ are the most common embedded programming languages. In contrast, Python raises the level of programming abstraction and programmer productivity. These are not mutually exclusive choices, however. PYNQ uses CPython which is written in C and integrates thousands of C libraries and can be extended with optimized code written in C. Wherever practical, the more productive Python environment should be used, and whenever efficiency dictates, lower-level C code can be used.

PYNQ is an open-source project that aims to work on any computing platform and operating system. This goal is achieved by adopting a web-based architecture, which is also browser agnostic. We incorporate the open-source Jupyter Notebook infrastructure to run an Interactive Python (IPython) kernel and a web server directly on the ARM processor of the Zynq device. The web server brokers access to the kernel via a suite of browser-based tools that provide a dashboard, bash terminal, code editors and Jupyter Notebooks. The browser tools are implemented with a combination of JavaScript, HTML and CSS and run on any modern browser [1].

PYNQ Architecture

PYNQ Overlays

The Xilinx Zynq All Programmable device is an SOC based on a dual-core ARM Cortex-A9 processor (referred to as the processing system or PS), integrated with FPGA fabric (referred to as programmable logic or PL). The PS subsystem includes a number of dedicated peripherals (memory controllers, USB, UART, IIC, SPI, etc.) and can be extended with additional hardware IP in a PL overlay (Fig. 1).

Overlays, or hardware libraries, are programmable/configurable FPGA designs that extend the user application from the processing system of the Zynq into the programmable logic. Overlays can be used to accelerate a software application or to customize the hardware platform for a particular application.

For example, image processing is a typical application where the FPGAs can provide acceleration. A software programmer can use an overlay in a similar way to a software library to run some of the image processing functions (e.g., edge detect, thresholding, etc.) on the FPGA fabric. Overlays can be loaded to the FPGA dynamically, as required, just like a software library. In this example, separate image

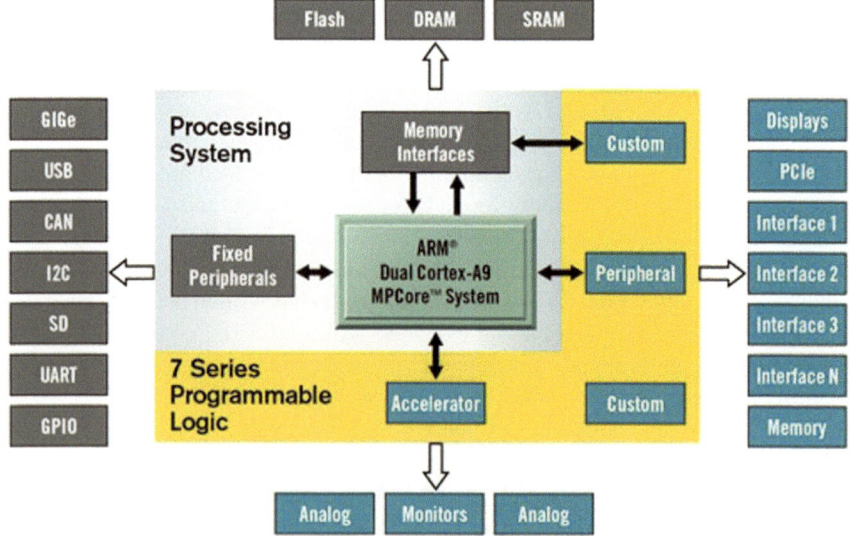

Fig. 1 PYNQ overlays

processing functions could be implemented in different overlays and loaded from Python on demand.

PYNQ provides a Python interface to allow overlays in the PL to be controlled from Python running in the PS. FPGA design is a specialized task which requires hardware engineering knowledge and expertise. PYNQ overlays are created by hardware designers and wrapped with this PYNQ Python API. Software developers can then use the Python interface to program and control specialized hardware overlays without needing to design an overlay themselves. This is analogous to software libraries created by expert developers which are then used by many other software developers working at the application level [2].

By default, an overlay (bitstream) called base is downloaded into the PL at boot time. The base overlay can be considered like a reference design for a board. New overlays can be installed or copied to the board and can be loaded into the PL as the system is running [3].

An overlay usually includes the following:

A bitstream to configure the FPGA fabric

A Vivado design HWH file to determine the available IP

Python API that exposes the IPs as attributes.

The PYNQ overlay class can be used to load an overlay. An overlay is instantiated by specifying the name of the bitstream file. Instantiating the overlay also downloads the bitstream by default and parses the HWH file.

Python code

```python
from pynq import Overlay
overlay = Overlay("base.bit")
```

For the base overlay, we can use the existing BaseOverlay class; this class exposes the IPs available on the bitstream as attributes of this class.

Python code

```python
from pynq.overlays.base import BaseOverlay
base_overlay = BaseOverlay("base.bit")
```

Once an overlay has been instantiated, the help() method can be used to discover what is in an overlay about. The help information can be used to interact with the overlay. Note that if you try the following code on your own board, you may see different results depending on the version of PYNQ you are using and which board you have.

Python code

```
help(base_overlay)
```

This will give a list of the IPs and methods available as part of the overlay.

From the help() print out above, it can be seen that in this case the overlay includes an leds instance, and from the report, this is an AxiGPIO class.

Running help() on the leds object will provide more information about the object including details of its API.

Python code

```
help(base_overlay.leds)
```

The API can be used to control the object. For example, the following cell will turn on LD0 on the board.

Python code

```
base_overlay.leds[0].toggle()
```

Information about other IPs can be found from the overlay instance in a similar way, as shown below.

The API can be used to control the object. For example, the following cell will turn on LD0 on the board.

Python code

```
help(base_overlay.video)
```

PYNQ overlays are analogous to software libraries. A programmer can download overlays into the Xilinx programmable logic at runtime to provide functionality required by the software application.

An overlay is a class of programmable logic design. Programmable logic designs are usually highly optimized for a specific task. Overlays, however, are designed to be configurable and reusable for broad set of applications. A PYNQ overlay will have a Python interface, allowing a software programmer to use it like any other Python package.

Jupyter Notebook

The material in this tutorial is specific to PYNQ. Wherever possible, however, it reuses generic documentation describing Jupyter Notebooks [4].

The Jupyter Notebook is an interactive computing environment that enables users to author notebook documents that include the following:

Live code
Interactive widgets
Plots
Narrative text
Equations
Images
Video.

The Jupyter Notebook combines three components as follows:

1. The notebook web application: An interactive web application for writing and running code interactively and authoring notebook documents.
2. Kernels: Separate processes started by the notebook web application that runs users' code in a given language and returns output back to the notebook web application. The kernel also handles things like computations for interactive widgets, tab completion and introspection.
3. Notebook documents: Self-contained documents that contain a representation of all content in the notebook web application, including inputs and outputs of the computations, narrative text, equations, images, and rich media representations of objects. Each notebook document has its own kernel.

Notebook Basics

The Notebook Dashboard

The notebook server runs on the ARM® processor of the board. You can open the notebook dashboard by navigating to pynq:9090 when your board is connected to the network. The dashboard serves as a home page for notebooks. Its main purpose is to display the notebooks and files in the current directory. For example, here is a screenshot of the dashboard page for an example directory:

The top of the notebook list displays clickable breadcrumbs of the current directory. By clicking on these breadcrumbs or on sub-directories in the notebook list, you can navigate your file system.

To create a new notebook, click on the "New" button at the top of the list and select a kernel from the dropdown (as seen below).

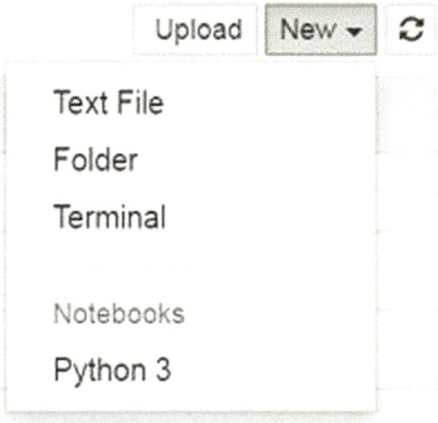

Notebooks and files can be uploaded to the current directory by dragging a notebook file onto the notebook list or by the "click here" text above the list.

The notebook list shows green "Running" text and a green notebook icon next to running notebooks (as seen below). Notebooks remain running until you explicitly shut them down; closing the notebook's page is not sufficient.

To shut down, delete, duplicate, or rename a notebook, check the checkbox next to it and an array of controls will appear at the top of the notebook list (as seen below). You can also use the same operations on directories and files when applicable.

To see all of your running notebooks along with their directories, click on the "Running" tab:

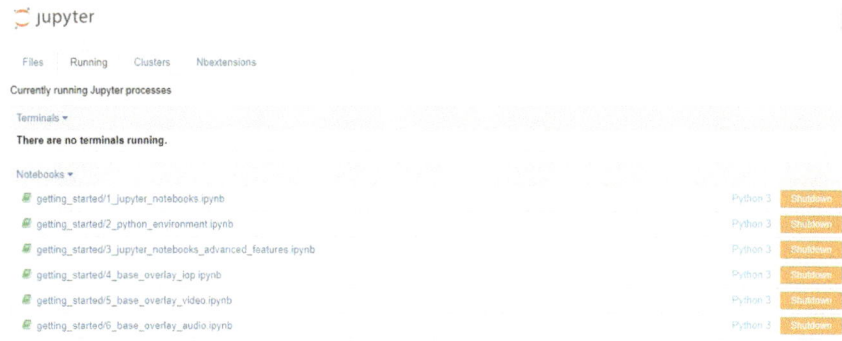

This view provides a convenient way to track notebooks that you start as you navigate the file system in a long running notebook server.

Overview of the Notebook UI

If you create a new notebook or open an existing one, you will be taken to the notebook user interface (UI). This UI allows you to run code and author notebook documents interactively. The notebook UI has the following main areas:
Menu
Toolbar
Notebook area and cells.
The notebook has an interactive tour of these elements that can be started in the "Help:User Interface Tour" menu item.

Modal Editor

The Jupyter Notebook has a modal user interface which means that the keyboard does different things depending on which mode the notebook is in. There are two modes: edit mode and command mode.

Edit Mode

Edit mode is indicated by a green cell border and a prompt showing in the editor area:

```
In [ ]: a=10
```

When a cell is in edit mode, you can type into the cell, like a normal text editor.
Enter edit mode by pressing "Enter" or using the mouse to click on a cell's editor area.

Command Mode

Command mode is indicated by a gray cell border with a blue left margin:

```
In [ ]: a=10
```

When you are in command mode, you are able to edit the notebook as a whole, but not type into individual cells. Most importantly, in command mode, the keyboard is mapped to a set of shortcuts that let you perform notebook and cell actions efficiently. For example, if you are in command mode and you press "c", you will copy the current cell—no modifier is needed.

Don't try to type into a cell in command mode; unexpected things will happen!

Enter command mode by pressing "Esc" or using the mouse to click outside a cell's editor area.

Mouse Navigation

All navigation and actions in the notebook are available using the mouse through the menubar and toolbar, both of which are above the main notebook area:

```
File    Edit    View    Insert    Cell    Kernel    Widgets    Help                    Trusted    | Python 3  O
```

Cells can be selected by clicking on them with the mouse. The currently selected cell gets a gray or green border depending on whether the notebook is in edit or command mode. If you click inside a cell's editor area, you will enter edit mode. If you click on the prompt or output area of a cell, you will enter command mode.

If you are running this notebook in a live session on the board, try selecting different cells and going between edit and command mode. Try typing into a cell.

If you want to run the code in a cell, you would select it and click the "play" button in the toolbar, the "Cell:Run" menu item or type Ctrl + Enter. Similarly, to copy a cell, you would select it and click the "copy" button in the toolbar or the "Edit:Copy" menu item. Ctrl + C, V are also supported.

Markdown and heading cells have one other state that can be modified with the mouse. These cells can either be rendered or unrendered. When they are rendered, you will see a nice formatted representation of the cell's contents. When they are unrendered, you will see the raw text source of the cell. To render the selected cell with the mouse, and execute it, click the "play" button in the toolbar or the "Cell:Run" menu item, or type Ctrl + Enter. To unrender the selected cell, double-click on the cell.

Keyboard Navigation

There are two different sets of keyboard shortcuts: one set that is active in edit mode and another in command mode.

The most important keyboard shortcuts are "Enter", which enters edit mode, and "Esc", which enters command mode.

In edit mode, most of the keyboard is dedicated to typing into the cell's editor. Thus, in edit mode, there are relatively few shortcuts. In command mode, the entire keyboard is available for shortcuts, so there are many more. The "Help"->"Keyboard Shortcuts" dialog lists the available shortcuts.

Some of the most useful shortcuts are as follows:

Basic navigation:"Enter","Shift-Enter","up/k","down/j"

Saving the notebook:"s"

Change cell types:"y","m","1–6","t"

Cell creation:"a","b"

Cell editing:"x","c","v","d","z"

Kernel operations:"i","0" (press twice).

Running Code

First and foremost, the Jupyter Notebook is an interactive environment for writing and running code. The notebook is capable of running code in a wide range of languages. However, each notebook is associated with a single kernel. PYNQ and this notebook are associated with the IPython kernel, which runs Python code.

Code Cells Allow You to Enter and Run Code

Run a code cell using "Shift-Enter" or pressing the "play" button in the toolbar above. The button displays run cell, select below when you hover over it.

```
In [1]: a=10

In [2]: print a
        10
```

There are two other keyboard shortcuts for running code:

"Alt-Enter" runs the current cell and inserts a new one below.

"Ctrl-Enter" runs the current cell and enters command mode.

Managing the Kernel

Code is run in a separate process called the kernel. The kernel can be interrupted or restarted. Try running the following cell and then hitting the "stop" button in the toolbar above. The button displays interrupt kernel when you hover over it.

```
In [ ]: import time
        time.sleep(10)
```

Cell Menu

The "Cell" menu has a number of menu items for running code in different ways. These includes the following:

Run and Select Below
Run and Insert Below
Run All
Run All Above
Run All Below.

Restarting the Kernels

The kernel maintains the state of a notebook's computations. You can reset this state by restarting the kernel. This is done from the menu bar or by clicking on the corresponding button in the toolbar.

sys.stdout

The stdout and stderr streams are displayed as text in the output area.

```
In [1]: print("Hello from Pynq!")
        Hello from Pynq!
```

Output Is Asynchronous

All output is displayed asynchronously as it is generated in the kernel. If you execute the next cell, you will see the output one piece at a time, not all at the end.

```
In [1]: import time, sys
        for i in range(8):
            print(i)
            time.sleep(0.5)
        0
        1
        2
        3
        4
        5
        6
        7
```

Large Outputs

To better handle large outputs, the output area can be collapsed. Run the following cell and then single- or double-click on the active area to the left of the output:

```
In [ ]: for i in range(50):
            print(i)
```

PYNQ Get Started

First, you need to ensure the integrity of the experimental equipment. Make sure you have the following five devices:

1. The PYNQ board
2. Power supply and power cord
3. Network cable
4. SD card
5. Card reader.

The above devices are necessary in subsequent experiments. Please keep them and don't lose them (Fig. 2).

Download the required software. From http://www.pynq.io/board.html to download the latest PYNQ-Z2 Boot Image V2.7.Exact the zip file. And you can get the image file (Fig. 3).

Prepare Win32DiskImager software. From https://sourceforge.net/projects/win32 diskimager/ to download win32diskimager (Fig. 4).

Write image to SD card. Plug SD card in card reader, then plug card reader in computer (Fig. 5).

Note: If the computer shows that the memory of 32G SD card is only more than 40M, it is because the image has been written before. Just write the image again directly.

Install and run Win32DiskImager. Next, choose the download img file. Click the "write" button and then wait for image file to be written to the SD card (Fig. 6).

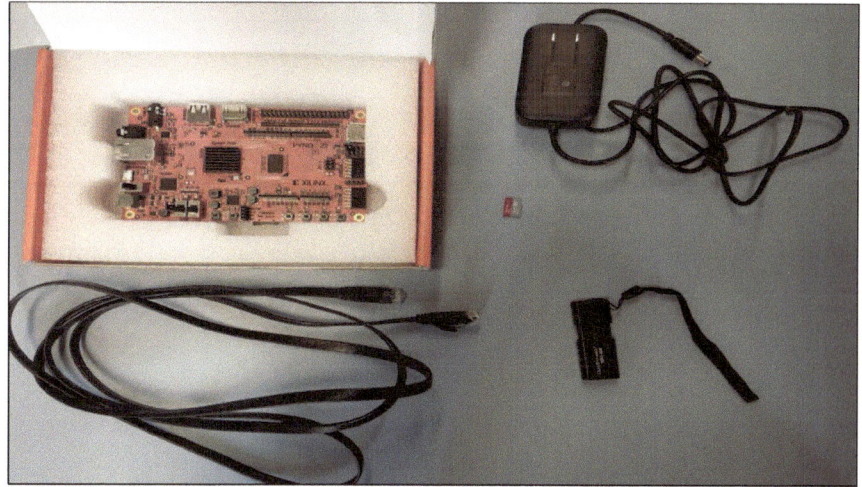

Fig. 2 Required equipments

Downloadable PYNQ images

If you have a Zynq board, you need a PYNQ SD card image to get started. You can download a pre-compiled PYNQ image from the table below. If an image is not available for your board, you can build your own SD card image (see details below).

Board	SD card image	Documentation	Vendor webpage
PYNQ-Z2	v2.7	PYNQ setup guide	TUL Pynq-Z2
PYNQ-Z1	v2.7	PYNQ setup guide	Digilent Pynq-Z1
PYNQ-ZU	v2.7	GitHub project page	TUL PYNQ-ZU
Kria KV260 Starter Kit*	Ubuntu image	Kria PYNQ setup	Xilinx Kria KV260
ZCU104	v2.7	PYNQ setup guide	Xilinx ZCU104
RFSoC 2x2	v2.7	RFSoC 2x2 GitHub Pages	XUP RFSoC 2x2
ZCU111	v2.7	PYNQ RFSoC workshop	Xilinx ZCU111
Ultra96V2	v2.7	Avnet PYNQ documentation	Avnet Ultra96V2
Ultra96 (legacy)	v2.7	See Ultra96V2	See Ultra96V2
TySOM-3-ZU7EV	v2.7	GitHub project page	Aldec TySOM-3-ZU7EV
TySOM-3A-ZU19EG	v2.7	GitHub project page	Aldec TySOM-3A-ZU19EG

Fig. 3 Download PYNQ image

Fig. 4 Download Win32 Disk Imager

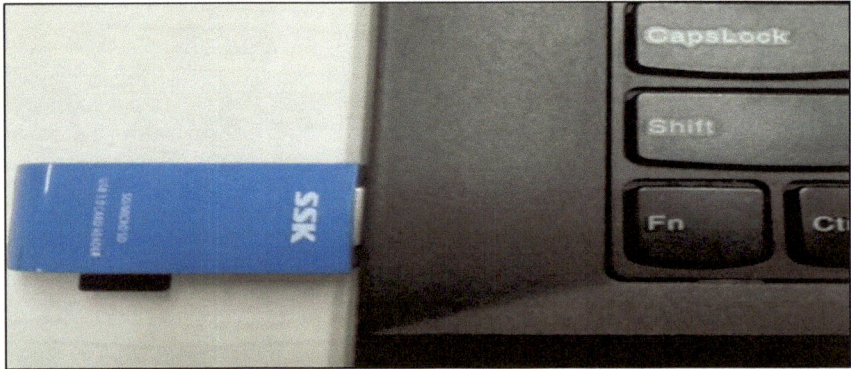

Fig. 5 Write image to SD card

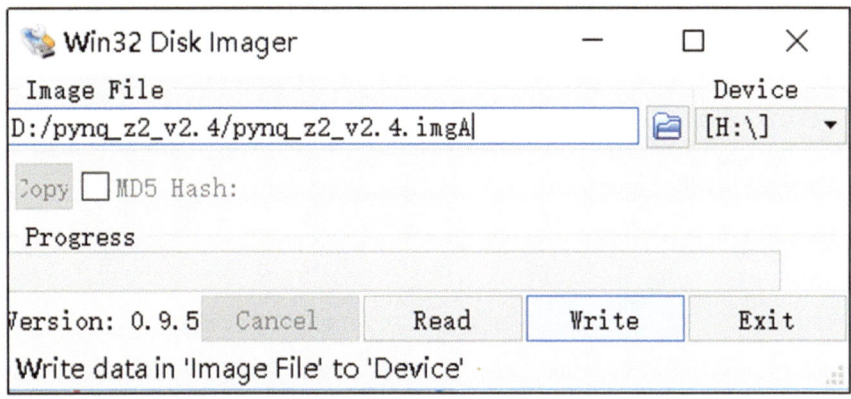

Fig. 6 Writing to the SD card

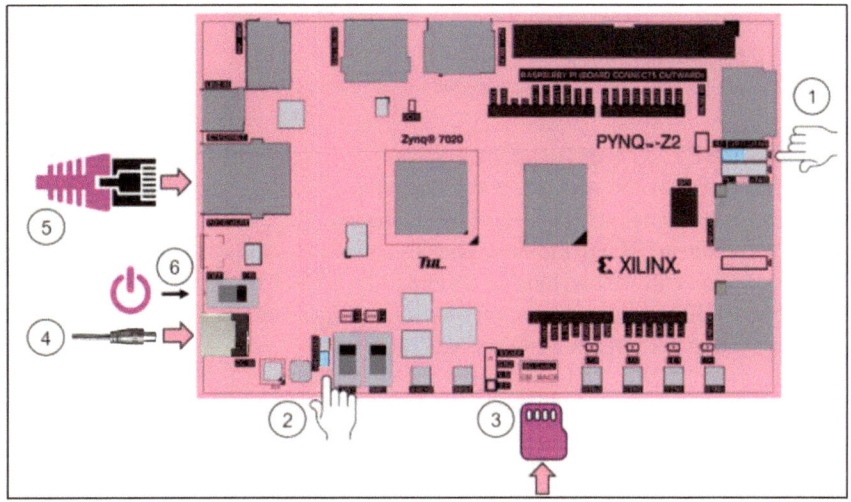

Fig. 7 Power line to power the PYNQ

Now start up your PYNQ. You can use power line to power the PYNQ (recommend) (Fig. 7) or use USB to power the PYNQ (not recommend) (Fig. 8).

Set the jumper for starting mode. There are three start modes for PYNQ: SD, OSPI and JTAG. To start the PYNQ using SD card, set the jumper to the SD position.

Set the jumper for power mode. There are two start modes for PYNQ: REG and USB. We recommend you to power your PYNQ using power line, so please set the jumper to the REG position. Or if you want to power your PYNQ using USB of your computer, please set the jumper to the USB position. Plug SD card in your PYNQ. Power the PYNQ. We recommend you to power your PYNQ using power line or if you want to power your PYNQ using USB of your computer (Fig. 9).

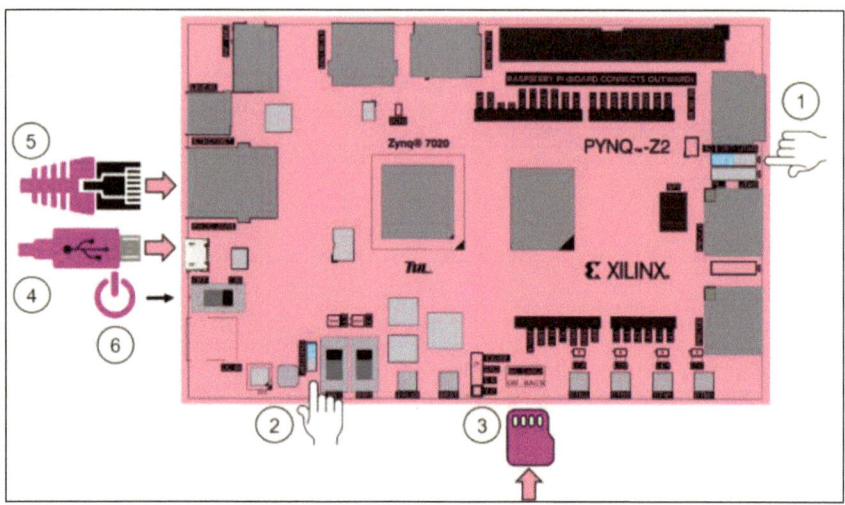

Fig. 8 USB to power the PYNQ

Connect the PYNQ board to a router or computer using a network cable. Depending on the choice of the network port (computer or router), subsequent operations will be different, which will be discussed in the next step.

Put the switch on ON. Put the switch on ON and wait for the system to start. About a minute later, you will see two blue LEDs and four yellow LEDs flash simultaneously, then the blue LEDs will be turned off and four yellow LEDs will be on, which will mean the system has already been started (Fig. 10).

Connect the PYNQ. You will need to have an Ethernet port available on your computer, and you will need to have permissions to configure your network interface. With a direct connection, you will be able to use PYNQ, but unless you can bridge the Ethernet connection to the board to an Internet connection on your computer, your board will not have Internet access. You will be unable to update or load new packages without Internet access.

1. Connect the computer and PYNQ directly with the Ethernet cable.
2. Assign your computer a static IP address on windows:

Open Network and Sharing Center. Click Ethernet. Choose properties. Select protocol vision 4 (TCP/IPv4) and then click the "properties button". Select the "Use the following IP address" option, and then type in the IP address :192.168.2.1, subnet mask:255.255.255.0, click OK. Browse to http://192.168.2.99:9090. Chrome is recommended. And you can see the interface as follows:

Type the password : xilinx . And now your access is successful (Fig. 11).

Only the network cable is used for Ethernet connection, which can smoothly use the Jupyter Notebook platform for subsequent operations. However, if you want to download or update relevant library files, you need to use PYNQ to connect to the Internet.

Fig. 9 Pluging SD card

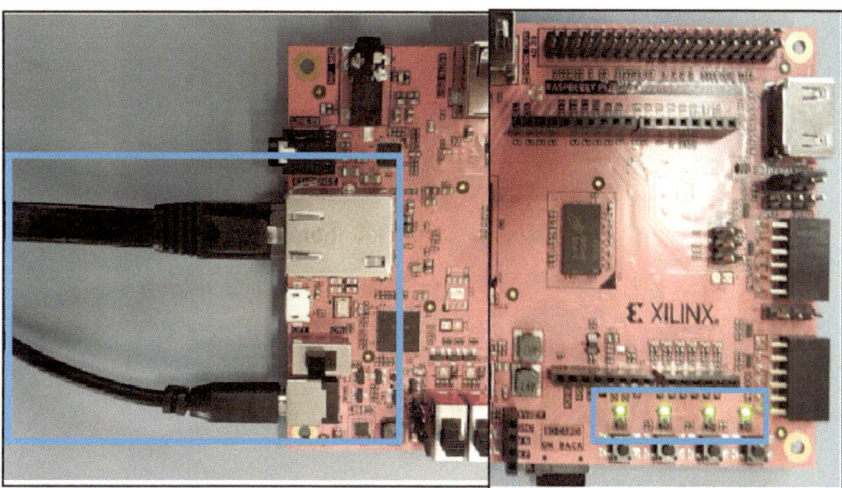

Fig. 10 Put the switch ON

Connect the power cord as shown in the figure and connect it to the computer with the network cable (Fig. 12).

Log in to the Jupyter Notebook terminal after connecting in the Ethernet mode described earlier.

Fig. 11 Interface

Fig. 12 Connect the power cord

Configure the gateway. Enter the command "sudo vi/ etc. / network / interfaces". Press "I" (insert) to edit after entering. Set the network address to 192.168.137.100 and the gateway to 192.168.137.1 (Fig. 13).

Linux code

```
auto eth0
iface eth0 inet static
address 192.168.137.100
netmask 255.255.225.0
gateway 192.168.137.1
source-directory/etc/network/interface.d
```

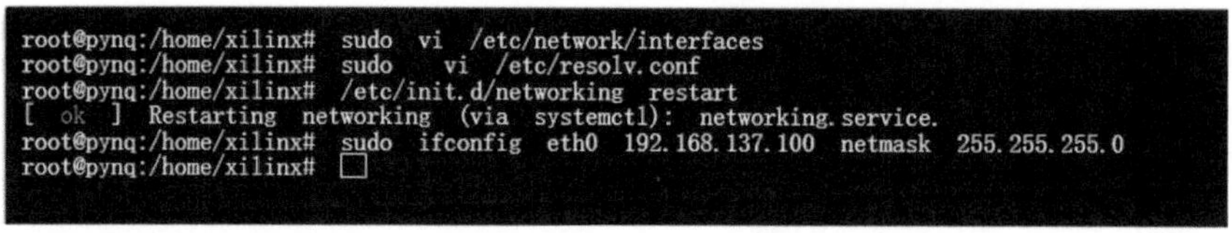

```
root@pynq:/home/xilinx# sudo vi /etc/network/interfaces
root@pynq:/home/xilinx# sudo    vi /etc/resolv.conf
root@pynq:/home/xilinx# /etc/init.d/networking restart
[ ok ] Restarting networking (via systemctl): networking.service.
root@pynq:/home/xilinx# sudo ifconfig eth0 192.168.137.100 netmask 255.255.255.0
root@pynq:/home/xilinx# ▢
```

Fig. 13 Set the serial port IP to 192.168.137.100

After editing, switch to English input method, press "Ctrl + C", and then press ": wq!" to exit editing. Then we continue to configure DNS. Enter the command "vi /etc/resolv.conf". Press "I" (insert) to edit after entering. Edit the command line as shown in the following code.

Linux code

```
namesever 127.0.0.53
namesever 114.114.114.114
namesever 114.114.115.115
search mshome.net
```

After editing, switch to English input method, press "Ctrl + C", and then press ": wq!" to exit editing. Restart the network. Set the serial port IP to 192.168.137.100.

On the computer, go to the network and sharing center in the control panel and change the adapter settings. Enter the network connection for network sharing: After sharing, the shared network IP is 192.168.137.1. After setting, Jupyter Notebook needs to re-enter 192.168.137.100. Conduct connection test and select any website (Fig. 14).

The website can be connected normally, and the Internet connection is successful.

Usually, we use the network cable to connect the computer segment, which is enough to complete the task of starting PYNQ for operation. When there is no network cable, we can also use serial port connection to enter the PYNQ operation interface (Fig. 15).

As shown in the figure, connect the USB cable to the UART serial port of PYNQ and connect it with the computer. From https://mobaxterm.mobatek.net/download. html, download mobaxterm. After downloading, open Mobaxterm and enter the new session (Fig. 16).

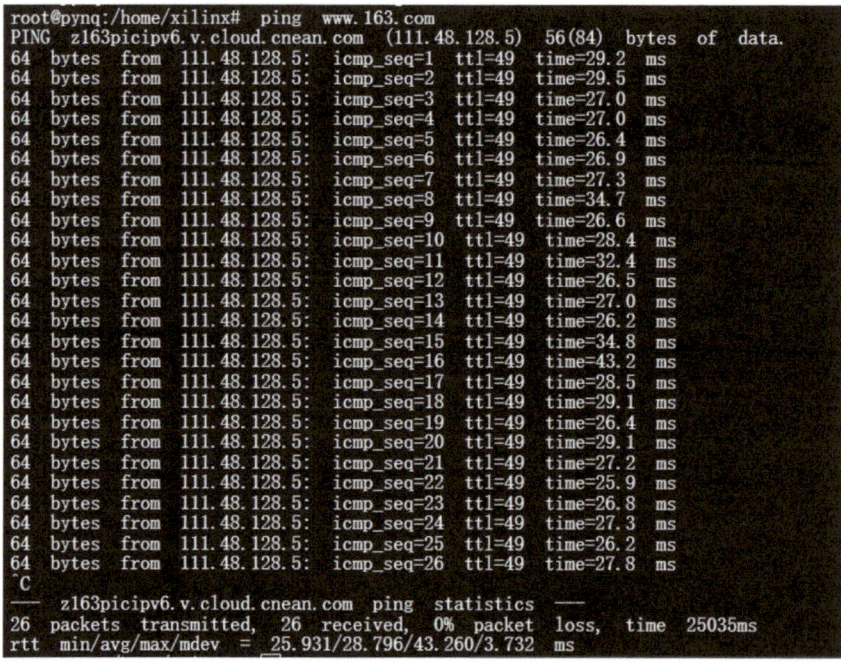

```
root@pynq:/home/xilinx# ping www.163.com
PING z163picipv6.v.cloud.cnean.com (111.48.128.5) 56(84) bytes of data.
64 bytes from 111.48.128.5: icmp_seq=1 ttl=49 time=29.2 ms
64 bytes from 111.48.128.5: icmp_seq=2 ttl=49 time=29.5 ms
64 bytes from 111.48.128.5: icmp_seq=3 ttl=49 time=27.0 ms
64 bytes from 111.48.128.5: icmp_seq=4 ttl=49 time=27.0 ms
64 bytes from 111.48.128.5: icmp_seq=5 ttl=49 time=26.4 ms
64 bytes from 111.48.128.5: icmp_seq=6 ttl=49 time=26.9 ms
64 bytes from 111.48.128.5: icmp_seq=7 ttl=49 time=27.3 ms
64 bytes from 111.48.128.5: icmp_seq=8 ttl=49 time=34.7 ms
64 bytes from 111.48.128.5: icmp_seq=9 ttl=49 time=26.6 ms
64 bytes from 111.48.128.5: icmp_seq=10 ttl=49 time=28.4 ms
64 bytes from 111.48.128.5: icmp_seq=11 ttl=49 time=32.4 ms
64 bytes from 111.48.128.5: icmp_seq=12 ttl=49 time=26.5 ms
64 bytes from 111.48.128.5: icmp_seq=13 ttl=49 time=27.0 ms
64 bytes from 111.48.128.5: icmp_seq=14 ttl=49 time=26.2 ms
64 bytes from 111.48.128.5: icmp_seq=15 ttl=49 time=34.8 ms
64 bytes from 111.48.128.5: icmp_seq=16 ttl=49 time=43.2 ms
64 bytes from 111.48.128.5: icmp_seq=17 ttl=49 time=28.5 ms
64 bytes from 111.48.128.5: icmp_seq=18 ttl=49 time=29.1 ms
64 bytes from 111.48.128.5: icmp_seq=19 ttl=49 time=26.4 ms
64 bytes from 111.48.128.5: icmp_seq=20 ttl=49 time=29.1 ms
64 bytes from 111.48.128.5: icmp_seq=21 ttl=49 time=27.2 ms
64 bytes from 111.48.128.5: icmp_seq=22 ttl=49 time=25.9 ms
64 bytes from 111.48.128.5: icmp_seq=23 ttl=49 time=26.8 ms
64 bytes from 111.48.128.5: icmp_seq=24 ttl=49 time=27.3 ms
64 bytes from 111.48.128.5: icmp_seq=25 ttl=49 time=26.2 ms
64 bytes from 111.48.128.5: icmp_seq=26 ttl=49 time=27.8 ms
^C
--- z163picipv6.v.cloud.cnean.com ping statistics ---
26 packets transmitted, 26 received, 0% packet loss, time 25035ms
rtt min/avg/max/mdev = 25.931/28.796/43.260/3.732 ms
```

Fig. 14 Internet connection test

Fig. 15 Use the network cable to connect the computer segment

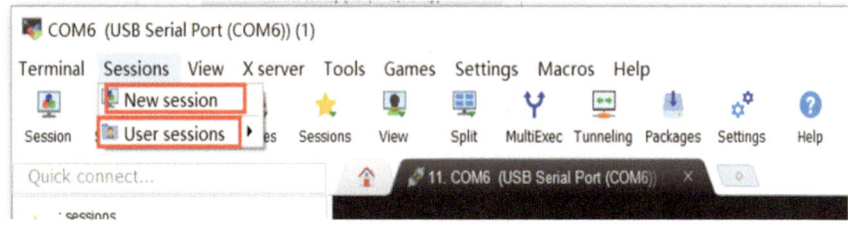

Fig. 16 Open Mobaxterm

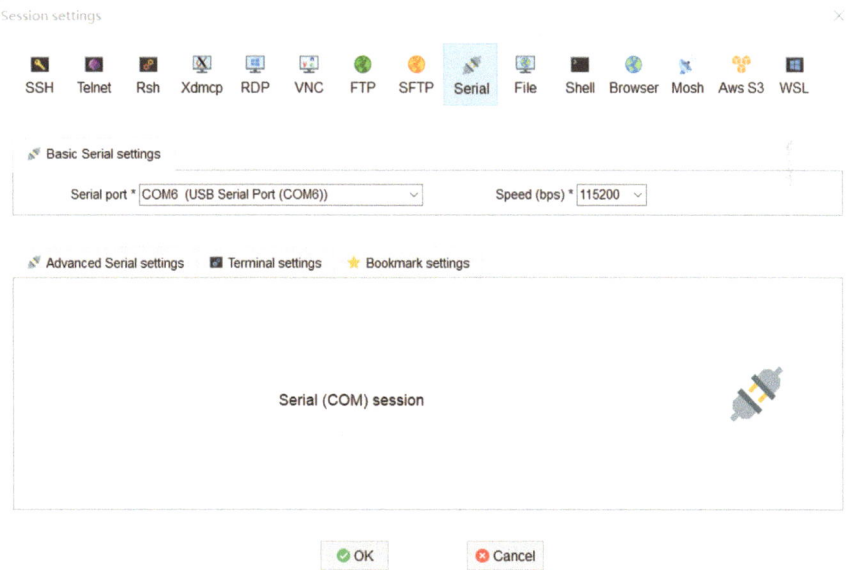

Fig. 17 Select the serial port option on the interface

Select the serial port option on the interface. Select the serial port option on the interface, select the displayed serial port, and set the baud rate to 115200 (Fig. 17).

Enter the Linux interface, as shown in Fig. 18. Try to query the IP address of the board and enter the command "ifconfig".

In this way, you can also carry out relevant experiments through serial port.

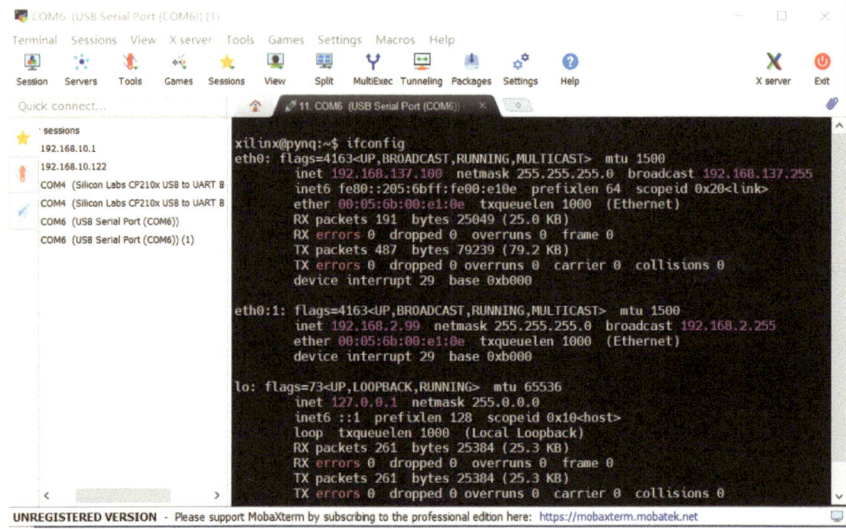

Fig. 18 Enter the Linux interface

References

1. Xilinx, Inc, PYNQ Introduction. https://pynq.readthedocs.io/en/latest/Cited23Jun2022
2. Xilinx, Inc, PYNQ Overlays. https://pynq.readthedocs.io/en/latest/Cited23Jun2022
3. Xilinx, Inc. Loading an Overlay. https://pynq.readthedocs.io/en/latest/Cited23Jun2022
4. Xilinx, Inc. Jupyter Notebooks. https://pynq.readthedocs.io/en/latest/Cited23Jun2022

PYNQ HelloWorld: Image Resizing

Abstract Image processing is a major application scenario of a series of embedded programming devices such as PYNQ. With the powerful computing function of its processor and programmable logic device, the usual image processing algorithm can consume less time than the algorithm running on the computer alone and can save a lot of computing resources. In this chapter, we will focus on image processing operations using PYNQ, that is, scaling an image [1]. We will compare the running time of the same algorithm using processor software alone on the PS side and using programmable devices on the PL side.This will more intuitively show the advantages of the hardware implementation algorithm.

Principle of Image Resizing Algorithm

We will first use Jupyter Notebook and Python on the PS side to realize the image resizing experiment. In this example, we will use the bilinear interpolation [2].

Tips

Bilinear interpolation, also known as bilinear interpolation. In mathematics, bilinear interpolation is a linear interpolation extension of the interpolation function with two variables. Its core idea is to perform linear interpolation in two directions respectively. As an interpolation algorithm in numerical analysis, bilinear interpolation is widely used in signal processing, digital image and video processing.

Image Resizing Using PS

Image resize using PIL library. PIL is the Python Imaging Library. This library provides extensive file format support, an efficient internal representation and fairly powerful image processing capabilities. The image module provides a class with the same name which is used to represent a PIL image. The module also provides a number of factory functions, including functions to load images from files and to create new images [3].

We import these library files. (1) PIL library to load and resize the image; (2) NumPy to store the pixel array of the image; (3) Matplotlib to show the image in the notebook.

Python code

```python
from scipy import signal, special
import numpy as np
import matplotlib.pyplot as plt
import matplotlib.ticker as mticker
from matplotlib.font_manager import FontProperties
```

We will load image from the SD card and create an image object. Then we create a NumPy array of the pixels and show the original image size. It may takes a while to render a large picture. For better visual effect, we double the size of the canvas. The following code only changes the display size, not the picture itself:

Python code

```python
canvas = plt.gcf()
size = canvas.get_size_inches()
canvas.set_size_inches(size*2)
old_width, old_height = original_image.size
print("Image size: {}x{} pixels.".format(old_width, old_height))
_ = plt.imshow(original_image)
```

The selected original image is 1920 × 1080 pixels in size and is displayed on the canvas at 2× magnification (Fig. 1).

```
canvas = plt.gcf()
size = canvas.get_size_inches()
canvas.set_size_inches(size*2)

old_w, old_h = org_img.size
print("Image size: {}x{} pixels".format(old_w, old_h))
_ = plt.imshow(org_img)
```

Image size: 1920x1080 pixels

Fig. 1 Original image

```
resized_img = org_img.resize((new_w, new_h), Image.BILINEAR)

canvas = plt.gcf()
size = canvas.get_size_inches()
canvas.set_size_inches(size*2)
print("Image size: {}x{} pixels.".format(new_w, new_h))
_ = plt.imshow(resized_img)
```

Image size: 960x540 pixels.

Fig. 2 Enlarged image

We will set image resize dimensions.We will use resize() method from the PIL library. We map multiple input pixels to a single output pixels to downscale the image. The Python Imaging Library provides different resampling filters. We will just choose the bilinear filter in our example. Pick one nearest pixel from the input image. Ignore all other input pixels (Fig. 2).

Python code

```
resize_factor = 2
new_width = int(old_width/resize_factor)
new_height = int(old_height/resize_factor)
resized_image = original_image.resize((new_width, new_height),
Image.BILINEAR)
print("Image_size:_{}x{}_pixels.".format(new_width, new_height))
_ = plt.imshow(resized_image)
%%timeit
resized_image = original_image.resize((new_width, new_height),
Image.BILINEAR)
```

Principle of Hardware Acceleration

This reference design illustrates how to run a resizer IP on the programmable logic (PL) using Jupyter Notebooks and Python [4].

Different from the experiment on PS side, the image resizing operation on PL side is mainly completed on programmable logic devices. After encapsulating the operation code into the corresponding IP core, we store the image data on the PS side and read the data through the AXI bus to complete the image zooming operation (Fig. 3).

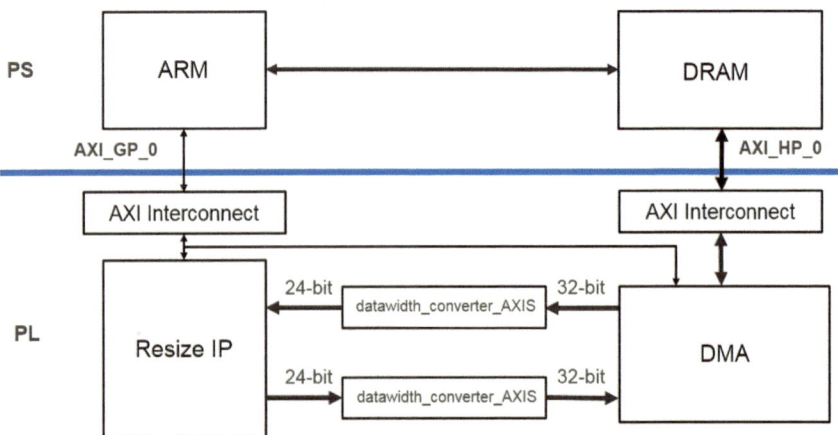

Fig. 3 PL architecture

Image Resizing Using PL

First, we import the library file.

Python code

```
from PIL import Image
import numpy as np
import matplotlib.pyplot as plt
%matplotlib inline
from pynq import Xlnk, Overlay
```

After the bitstream has been downloaded, the PL will be populated with the resize IP, the DMA engine and a few other components. The resize IP is configured to use bilinear interpolation and then create DMA and resizer IP objects.

Python code

```
resize_design = Overlay("resizer.bit")
dma = resize_design.axi_dma_0
resizer = resize_design.resize_accel_0
```

We will load image from the SD card and create a PIL image object. Let's also check the original image size. It may take a while to render a large picture. For better visual effect, we double the size of the canvas. The following code only changes the display size, not the picture itself:

Python code

```
canvas = plt.gcf()
size = canvas.get_size_inches()
canvas.set_size_inches(size*2)
old_width, old_height = original_image.size
print("Image_size:_{}x{}_pixels."
.format(old_width, old_height))
_ = plt.imshow(original_image)
```

```
canvas = plt.gcf()
size = canvas.get_size_inches()
canvas.set_size_inches(size+2)

old_w, old_h = ori_img.size
print("Image size: {}x{} pixels.".format(old_w, old_h))
_ = plt.imshow(ori_img)
```

Image size: 1920x1080 pixels.

Fig. 4 Original image

We can set resize dimensions. We now allocate memory to process data on PL. Data is provided as contiguous memory blocks. The size of buffer depends on the size of the input or output data. The image dimensions extracted from the read image are used to allocate contiguous memory blocks. We will call cma array() to perform the allocation (Fig. 4).

Python code

```
xlnk = Xlnk()
in_buffer = xlnk.cma_array(shape
=(old_height, old_width, 3),
 dtype=np.uint8, cacheable=1)
out_buffer = xlnk.cma_array(shape
=(new_height, new_width, 3),
 dtype=np.uint8, cacheable=1)
```

Tips

Documentation snippet for xlnk.cma array : Get a contiguously allocated numpy array.
Parameters
shape : int or tuple of int

The dimensions of the array to construct - We use (height, width, depth) dtype :
numpy.dtype or str
The data type to construct - We use 8-bit unsigned int

Note that the original image has to be copied into the contiguous memory array
(deep copy). We can now run the resizer IP. We will push the data from input buffer
through the pipeline to the output buffer. For ease of use, we will define a run kernel
function that will be called multiple times.

Python code

```python
def run_kernel():
 dma.sendchannel.transfer(in_buffer)
 dma.recvchannel.transfer(out_buffer)
 resizer.write(0x00,0x81) # start
 dma.sendchannel.wait()
 dma.recvchannel.wait()
```

We will also need to set up resizer and DMA IPs using MMIO interface before
we stream image data to them. For example, we need to write dimensions data to
MMIO registers of resizer. These register writings only have to be done once.

Tips

register offset configuration:
0x10 number of rows for original picture
0x18 number of columns for original picture
0x20 number of rows for resized picture
0x28 number of columns for resized picture

Python code

```python
def run_kernel():
 dma.sendchannel.transfer(in_buffer)
 dma.recvchannel.transfer(out_buffer)
 resizer.write(0x00,0x81)
 dma.sendchannel.wait()
 dma.recvchannel.wait()
```

```
canvas = plt.gcf()
size = canvas.get_size_inches()
canvas.set_size_inches(size*2)

print("Image size: {}x{} pixels.".format(new_w, new_h))
_ = plt.imshow(resized_img)

Image size: 960x540 pixels.
```

Fig. 5 Enlarged image

Now we can perform the resizing operation. We can also time the resize in PL operation. Finally, we need to reset all the contiguous memory buffers (Fig. 5).

Python code

```
run_kernel()
resized_image = Image.fromarray(out_buffer)
print("Image_size:_{}x{}_pixels."
.format(new_width, new_height))
_ = plt.imshow(resized_image)
%%timeit
run_kernel()
resized_image = Image.fromarray(out_buffer)
xlnk.xlnk_reset()
```

References

1. L. Crockett, D. Northcote, C. Ramsay, F. Robinson , R. Stewart, Exploring Zynq MPSoC: With PYNQ and machine learning applications (2019)
2. https://github.com/Xilinx/PYNQ/Cited21Jun2022
3. https://www.xilinx.com/support/documentation/sw_manuals/xilinx2017_1/ug1233-xilinx-opencv-user-guide.pdf/Cited21Jun2022
4. https://pillow.readthedocs.io/en/latest/handbook/concepts.htmlfilters/Cited22Jun2022

Digital Signal Processing Experiment on PYNQ

Abstract Speech signal processing is mainly a process of collecting the electrical signal converted by speech waveform through microphone and performing a series of mathematical operations to achieve related purposes. In the face of the impact of adverse factors such as noise, interference, acoustic echo and reverberation, signal processing, machine learning and other means are used to improve the signal-to-noise ratio or subjective auditory perception of the target speech and enhance the robustness of the subsequent links of speech interaction [1]. This chapter mainly introduces the speech signal processing experiment based on PYNQ. First, we receive the speech signal and analyze its characteristics to observe its time–frequency properties. Then, facing the noise of the original signal, we choose to use a filter for denoising. We introduce several commonly used filters. Finally, we carry out a comprehensive speech processing experiment on the signal.

Time Frequency Analysis of Audio Signal

This experiment uses the recording and playing function of PYNQ. After recording audio, we use the operation of time–frequency conversion to observe the spectrum information of recorded audio [2].

First, we write the code for recording audio. We first call the base bitstream file in PYNQ's overlay hardware library.

Python code

```
from pynq.overlays.base import BaseOverlay
vedio = BaseOverlay("base.bit")
pAudio = vedio.audio
```

Record a 3-second audio and store it as an audio file in PDM format. We then import the stored audio file and use the play command.

S. Sun et al. (eds.), *Experience of PYNQ*, SpringerBriefs in Applied Sciences and Technology, https://doi.org/10.1007/978-981-19-9072-4_3

Python code

```
pAudio.record(3)
  pAudio.save("Recording_1.pdm")
  pAudio.load("/home/xilinx/jupyter_notebooks/Recording_1.pdm")
  pAudio.play()
```

Now that we have a complete piece of audio data, and we will enter the data preprocessing stage of audio files. Our existing PDM signal is the information of one sampling point every 32 bits. In this step, we first convert the 32-bit integer buffer to 16-bit. Then we divide 16-bit words (16 1-bit samples each) into 8-bit words with 1-bit sample each.

Python code

```
import time
import numpy as np
start = time.time()
af_uint8 = np.unpackbits(pAudio.buffer.astype(np.int16)
.byteswap(True).view(np.uint8))
print("Time to convert {:,d} PDM samples: {:0.2f} seconds"
.format(np.size(pAudio.buffer)*16, end-start))
print("Size of audio data: {:,d} Bytes"
.format(af_uint8.nbytes))
```

Tips

PDM (Pulse Density Modulation) is a modulation method that uses binary numbers 0,1 to represent analog signals. In PDM signal, the amplitude of analog signal is expressed by the density of the corresponding area of output pulse.

Inside the PDM output microphone, a small ADC IC (modulator) can be found, which is used to convert the analog signal output by the sensor into PDM signal stream. The low-frequency spectrum of the signal generated by this audio converter technology is close to the required audio signal, and the frequency of the high-frequency parasitic part rises rapidly with the increase of the frequency above a certain inflection point frequency, which basically falls outside the audio frequency range required for the final product.

For each sampling point, 1 bit can be used to record, that is, only "0" indicating "no" and "1" indicating "yes" are used to record the level value of this sampling point.

The sampling interval of audio in PDM format is still too small. We need to convert the audio data from PDM format into PCM format by sampling. At this time, the sampling rate of audio data will be reduced from 3 MHz to 32 kHz.

Python code

```
import time
from scipy import signal
start = time.time()
af_dec = signal.decimate(af_uint8,8,zero_phase=True)
af_dec = signal.decimate(af_dec,6,zero_phase=True)
af_dec = signal.decimate(af_dec,2,zero_phase=True)
af_dec = (af_dec[10:-10]-af_dec[10:-10].mean())
end = time.time()
print("Time to convert {:,d} Bytes: {:0.2f} seconds"
.format(af_uint8.nbytes, end-start))
print("Size of audio data: {:,d} Bytes"
.format(af_dec.nbytes))
del af_uint8
```

Tips

PCM (Pulse Code Modulation) is to transform a time-continuous analog signal into a time-discrete digital signal, which is transmitted in the channel. Pulse code modulation is the process of sampling the analog signal first, then quantifying the amplitude of the sample value and coding.

Sampling is to scan the analog signal periodically and turn the continuous signal into a discrete signal in time. Sampling must follow Nyquist sampling theorem. After sampling, the analog signal should also contain all the information in the original signal, that is, it can recover the original analog signal without distortion. The lower limit of its sampling rate is determined by the sampling theorem.

Therefore, we have obtained the audio signal in PCM data format through sampling, which is very important for our subsequent digital signal processing operation.

? Questions

What is the difference between PDM and PCM?

Next, we draw the time domain analysis diagram of audio signal. PYNQ contains some basic model libraries of Python language, one of which is Matplotlib, which has

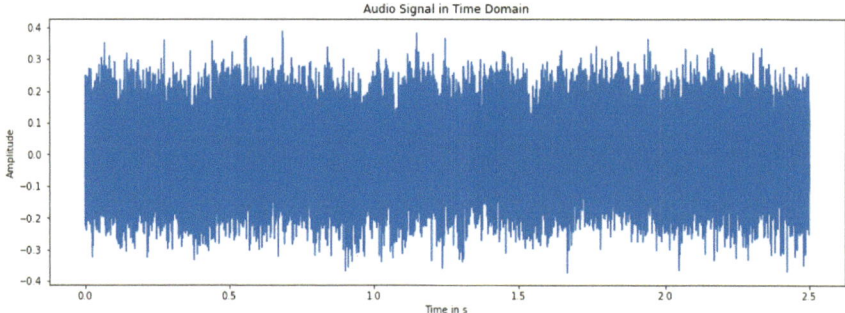

Fig. 1 Audio time domain graphic

a powerful drawing function. We call the Matplotlib library file and set the horizontal axis as the time axis and the vertical axis as the amplitude axis (note that the time is the data length divided by the sampling rate). In this way, we get the audio time domain graphics.

Python code

```
import numpy as np
import matplotlib.pyplot as plt
plt.figure(num=None, figsize=(15, 5))
time_axis = np.arange(0,((len(af_dec))/32000),1/32000)
plt.title('Audio Signal in Time Domain')
plt.xlabel('Time in s')
plt.ylabel('Amplitude')
plt.plot(time_axis, af_dec)
plt.show()
```

After getting the time domain diagram, we convert it to the frequency domain. PYNQ's own Python SciPy library has FFT transformation functions. Call FFT function for fast Fourier transform and draw a spectrum diagram (Fig. 1).

Python code

```
from scipy.fftpack import fft
yf = fft(af_dec)
yf_2 = yf[1:len(yf)//2]
xf = np.linspace(0.0, 32000//2, len(yf_2))
plt.figure(num=None, figsize=(15, 5))
plt.plot(xf, abs(yf_2))
```

```
plt.title('Magnitudes of Audio Signal Frequency Components')
plt.xlabel('Frequency in Hz')
plt.ylabel('Magnitude')
plt.show()
```

Design of a FIR Filter

The original audio signal has a lot of noise. From the analysis of the frequency domain, the noise is mostly in the high-frequency part, and the useful sound signal is mostly in the low-frequency part. Therefore, we need to eliminate the noise signal in the high-frequency part, and the filter is one of the methods to eliminate the noise [3].

We will implement an FIR filter on PYNQ (Fig. 2).

Tips

FIR (Finite Impulse Response) filter: finite length unit impulse response filter, also known as a non-recursive filter, is the most basic element in the digital signal processing system. It can ensure arbitrary amplitude-frequency characteristics and strict linear phase frequency characteristics. At the same time, its unit sampling response is finite, so the filter is a stable system. Therefore, FIR filter is widely used in communication, image processing, pattern recognition, and other fields.

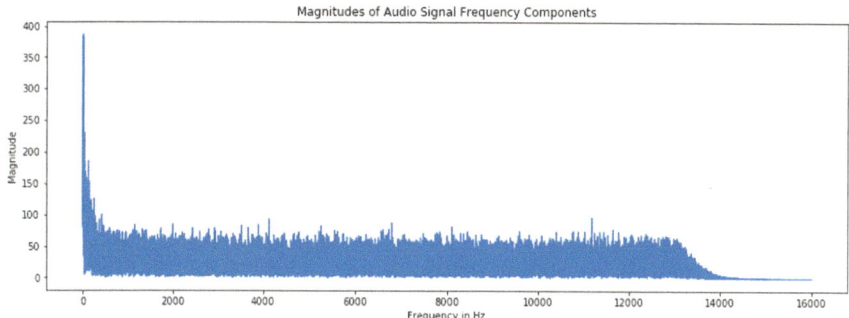

Fig. 2 Spectrum diagram

Construct FIR low-pass filter function using Python language [4].

Python code

```
from scipy import signal
def low_pass_FIR(data):
b = signal.firwin(15, 0.125)
low_data = signal.lfilter(b, 1, data)
return low_data
```

The firwin function uses the window method to design the FIR filter. This function calculates the coefficients of the finite impulse response filter. The filter will have a linear phase, type I if numtaps is odd and type II if numtaps i even.

The type II filter always has a zero response at the Nyquist frequency, so if you use numtaps to call firwin even and have a passband with the right end at the Nyquist frequency, you will throw a ValueError exception. This function calculates the coefficients of the finite impulse response filter. The filter will have a linear phase; type I if numtaps is odd and type II if numtaps is even.

The type II filter always has a zero response at the Nyquist frequency, so if you use numtaps to call firwin even and have a passband with the right end at the Nyquist frequency, you will throw a ValueError exception.

FIR needs to design the window function, and there are usually several commonly used window functions for design:

1. Rectangular window. Rectangular window belongs to the zero power window of time variable,which is defined as

$$\omega(n) = R_M(n) = \begin{cases} 1, & 0 \le n \le M - 1 \\ 0, & \text{else} \end{cases} \tag{1}$$

   Rectangular windows are used the most. It is customary not to add windows to make the signal pass through the rectangular windows. The advantage of this window is that the main lobe is relatively concentrated, but the disadvantage is that the side lobe is high and has a negative side lobe, which leads to high-frequency interference and leakage in the transformation and even negative spectrum phenomenon.

2. Triangular window, also known as Fejer window, is a power window. Compared with rectangular window, the width of main lobe is about twice that of rectangular window, but the side lobe is small and there is no negative side lobe.

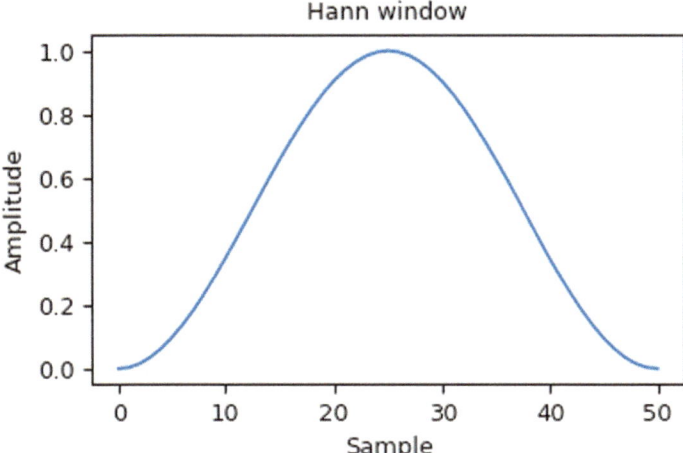

Fig. 3 Hanning window

3. The Hamming window. The Hamming window is defined as

$$w(n) = 0.54 - 0.46\cos\left(\frac{2\pi n}{M-1}\right) \quad 0 \le n \le M-1 \qquad (2)$$

The Hamming was named for R. W. Hamming, an associate of J. W. Tukey and is described in Blackman and Tukey. It was recommended for smoothing the truncated autocovariance function in the time domain. Most references to the Hamming window come from the signal processing literature, where it is used as one of many windowing functions for smoothing values. It is also known as an apodization (which means "removing the foot", i.e., smoothing discontinuities at the beginning and end of the sampled signal) or tapering function.

4. Hanning window. Hanning window is also known as rising cosine window. Hanning window can be regarded as the sum of the spectrum of three rectangular time windows, or the sum of three $sinc(T)$ functions. The two terms in brackets move π / T to the left and right relative to the first spectral window, so that the side lobes cancel each other and eliminate high-frequency interference and energy leakage. It can be seen that the main lobe of Hanning window is widened and reduced, while the side lobe is significantly reduced. From the point of view of reducing leakage, Hanning window is better than rectangular window. However, the widening of main lobe of Hanning window is equivalent to the widening of analysis bandwidth and the decline of frequency resolution (Fig. 3).

References

1. R.A. Roberts, C.T. Mullis (1987) Digital Signal Processing. Addison-Wesley Longman Publishing Co., Inc
2. https://github.com/Xilinx/PYNQ/blob/v2.0/boards/Pynq-Z1/base/notebooks/arduino/arduino_grove_ledbar.ipynb
3. https://numpy.org/doc/stable/reference/routines.window.html
4. https://www.fpgadeveloper.com/2018/03/how-to-accelerate-a-python-function-with-pynq.html/

Basic Communication Experiment on PYNQ

Abstract Communication technology and communication industry have been one of the fastest-growing fields since the 1980s. This is true both internationally and domestically. This is one of the important signs that mankind has entered the information society. Communication is the exchange of information. In this sense, communication has existed in ancient times. Digital communication is the communication of transmitting digital signals. The analog signals sent by the source are encoded into digital signals by the desire of the digital terminal. The digital signals sent by the terminal are encoded into digital signals suitable for channel transmission, and then the signals are modulated to the digital channels used by the system by the modem. After transmission to the opposite segment, they are finally transmitted to the destination through reverse transformation. Digital communication has become the most important basis of communication technology in modern communication networks because of its strong anti-interference ability, easy storage, processing and exchange and is widely used in various communication systems of modern communication networks [1]. In this chapter, we focus on the realization of communication function on PYNQ. We first introduce several common modulation and demodulation methods in modern communication. Then we are going to implement a classic modern communication system, namely orthogonal frequency division multiplexing (OFDM)

Modulation and Demodulation (BPSK, QPSK, QAM)

Modulation is the process of processing the information of the signal source and adding it to the carrier to make it suitable for channel transmission, which is the technology of changing the carrier with the signal. Generally speaking, the information of signal source (also known as signal source) contains DC component and frequency component with lower frequency, which is called baseband signal. Baseband signals are often not used as transmission signals, so baseband signals must be converted into a signal with a very high frequency relative to baseband frequency to be suitable for channel transmission. This signal is called modulated signal, and baseband signal is called modulated signal. Modulation is realized by changing the

S. Sun et al. (eds.), *Experience of PYNQ*, SpringerBriefs in Applied Sciences and Technology, https://doi.org/10.1007/978-981-19-9072-4_4

amplitude, phase or frequency of the carrier signal of the high-frequency carrier, that is, the message, so that it changes with the amplitude of the baseband signal. Demodulation is the process of extracting the baseband signal from the carrier for the predetermined receiver (also known as the sink) to process and understand.

Let's first analyze the modulation and demodulation process of QPSK:

The code realizes QPSK modulation by using IQ modulation principle [2]. The digital baseband signal is divided into I and Q channels, which are multiplied by the carrier wave, respectively, and finally added to obtain the modulated signal. The difficulty in the implementation process is how to realize the polarity conversion of the signal and realize the serial parallel conversion to separate the odd and even bits [3]. First, we import the relevant library functions.

Python code

```
from scipy import signal, special
import numpy as np
import matplotlib.pyplot as plt
import matplotlib.ticker as mticker
from matplotlib.font_manager import FontProperties
```

Next, we define the baseband signal. The parameters of baseband signal include baseband signal frequency, transmitted bits, sampling interval, sampling frequency, carrier frequency and signal-to-noise ratio.

Python code

```
T = 1
nb = 100
delta_T = T/200
fs = 1/delta_T
fc = 10/T
SNR = 0
```

Call the random function to generate any 1*nb matrix from 0 to 1. If it is greater than 0.5, it will be displayed as 1, and if it is less than 0.5, it will be displayed as 0. Create a zero matrix to transform the baseband signal into the corresponding waveform signal.

Python code

```
data = [1 if x > 0.5 else 0 for x in np.random.randn(1, nb)[0]]
data0 = []
for q in range(nb):
    data0 += [data[q]]*int(1/delta_T)
```

Generate modulated signal.

Python code

```
data1 = []
datanrz = np.array(data)*2-1
for q in range(nb):
    data1 += [datanrz[q]]*int(1/delta_T)

idata = datanrz[0:(nb-1):2]
qdata = datanrz[1:nb:2]
ich = []
qch = []
for i in range(int(nb/2)):
    ich += [idata[i]]*int(1/delta_T)
    qch += [qdata[i]]*int(1/delta_T)

a = []
b = []
for j in range(int(N/2)):
    a.append(np.math.sqrt(2/T)*np.math.cos(2*np.math.pi*fc*t[j]))
    b.append(np.math.sqrt(2/T)*np.math.sin(2*np.math.pi*fc*t[j]))
idata1 = np.array(ich)*np.array(a)
qdata1 = np.array(qch)*np.array(b)
s = idata1 + qdata1
```

So far, the generation of QPSK modulation signal has been realized by code. You can observe the waveform of the modulation signal by adding the code of the drawing. Now we continue the demodulation process of QPSK. Suppose our signal passes through Gaussian channel, demodulates and obtains the original signal through low-pass filter.

Python code

```
s11 = wgn(s, SNR)
s1 = s + s11
idata2 = s1*np.array(a)
qdata2 = s1*np.array(b)
[b, a] = signal.butter(2, 2*fc/fs)
idata22 = signal.filtfilt(b, a, idata2)
qdata22 = signal.filtfilt(b, a, qdata2)
demodata0 = idata22 + qdata22
```

Tips

The Gaussian channel is a radio frequency communication channel, which contains the characteristics of specific noise spectral density at various frequencies, resulting in arbitrary distribution of errors in the channel.

QPSK is one of the classic signal modulation methods. The specific modulation principles of other signal modulation methods will not be further elaborated. We can use the Python library scikit CommPy to implement a variety of signal modulation methods.

Python code

```
import commpy as cpy
bits = np.random.binomial(n=1,p=0.5,size=(128))
Modulation_type ="BPSK"
if Modulation_type=="BPSK":
    bpsk = cpy.PSKModem(2)
    symbol = bpsk.modulate(bits)
    return symbol
elif Modulation_type=="QPSK":
    qpsk = cpy.PSKModem(4)
    symbol = qpsk.modulate(bits)
    return symbol
elif Modulation_type=="8PSK":
    psk8 = cpy.PSKModem(8)
    symbol = psk8.modulate(bits)
    return symbol
```

```
elif Modulation_type=="16QAM":
    qam16 = cpy.QAMModem(16)
    symbol = qam16.modulate(bits)
    return symbol
elif Modulation_type=="64QAM":
    qam64 = cpy.QAMModem(64)
    symbol = qam64.modulate(bits)
    return symbol
```

Comprehensive Experiment—Design of OFDM Communication System

Orthogonal frequency division multiplexing (OFDM) is the orthogonal frequency division multiplexing technology. In fact, OFDM is multi-carrier modulation (MCM), a kind of multi-carrier modulation. The parallel transmission of high-speed serial data is realized by frequency division multiplexing. It has a good anti-multipath fading ability and can support multi-user access [4].

In a communication system, the bandwidth provided by a channel is usually much wider than that required to transmit a signal. If a channel transmits only one signal, it is very wasteful. To make full use of the bandwidth of the channel, the method of frequency division multiplexing can be adopted.

The main idea of OFDM is to divide the channel into several orthogonal subchannels, convert the high-speed data signal into parallel low-speed sub-data streams and modulate them to transmit on each subchannel [5]. Orthogonal signals can be separated by using correlation technology at the receiving end, which can reduce the mutual interference (ISI) between subchannels. The signal bandwidth on each subchannel is smaller than the relevant bandwidth of the channel, so each subchannel can be seen as flat fading, so intersymbol crosstalk can be eliminated. Moreover, since the bandwidth of each subchannel is only a small part of the original channel bandwidth, channel equalization becomes relatively easy.

References

1. A.C. Partridge, Principles of communication. Mine Quarry; (United States) 7(7/8) (1978)
2. S. Gronemeyer, A. McBride, MSK and offset QPSK modulation. IEEE Trans. Commun. 24(8), 809–820 (1976)
3. www.rfsoc-pynq.io/overlays
4. H. Bolcskei, MIMO-OFDM wireless systems: basics, perspectives, and challenges. IEEE Wirel Commun 13(4), 31–37 (2006)
5. Y.S. Cho, J. Kim, W.Y. Yang, et al., MIMO-OFDM wireless communications with MATLAB (2010)

Neural Network Experiment on PYNQ

Abstract This chapter presents modifiable frameworks for fast and easy neural network prototyping on the Xilinx PYNQ platform. With a Python-based programming interface, the framework combines the convenience of high-level abstraction with the speed of optimized FPGA implementation. We will introduce the theory of PYNQ acceleration through the following two experiments.

MNIST Handwritten Numeral Recognition Based on CNN

The experimental system flowchart and block diagram of MNIST handwritten numerical recognition based on CNN are shown in Figs. 1 and 2.

Experimental development environment

Python 3.6.5; Vivado HLS 2016.1; Vivado 2016.1;

Software Section

Fully Connected Neural Network

Fully connected neural network is mainly for image recognition. In the human eye, the picture is colored, and in the computer, the color of the picture is represented by a gray value between 0 and 1. The simplest black and white photo is to use 0 for white, 1 for black, and the median value for gray. The color picture is represented by a three-channel mix of red, green and blue (RGB). The traditional handwritten numeral recognition idea is that for each pixel in the picture (i.e., the point at each position in the matrix), there is a support rate corresponding to different classification results (0–9). And then the support rate for each classification in the image is added

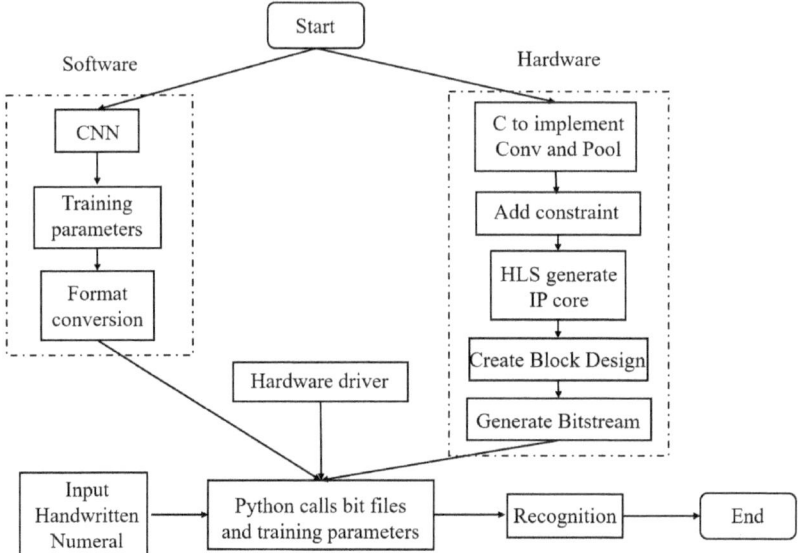

Fig. 1 System flow chart

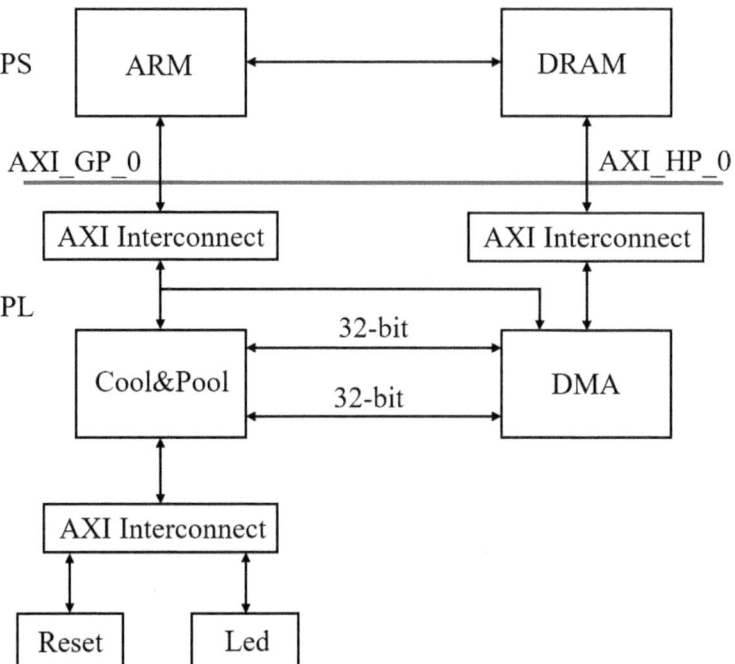

Fig. 2 System block diagram

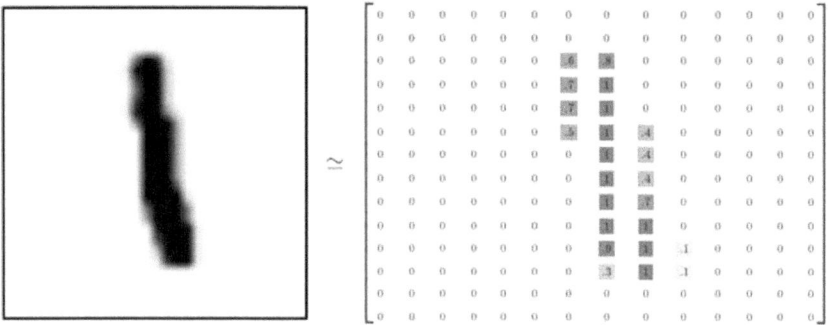

Fig. 3 Convert gray image into two-dimensional matrix

up, and the classification with the highest support rate is used as the recognition result of this picture, that is, y = softmax (wx + b). (Where softmax normalizes the calculation results to facilitate backpropagation calculations during training) Fig. 3.

Python code

```
#Main code of num_recognition.py
mnist = input_data.read_data_sets("MNIST_data/", one_hot=True)

x = tf.placeholder(tf.float32, [None, 784])
W = tf.Variable(tf.zeros([784, 10]))
b = tf.Variable(tf.zeros([10]))
y = tf.nn.softmax(tf.matmul(x, W) + b)
y_ = tf.placeholder(tf.float32, [None, 10])
cross_entropy = tf.reduce_mean(−tf.reduce_sum(y_ * tf.log(y),
                            reduction_indices=[1]))
train_step = tf.train.GradientDescentOptimizer(0.5).
                        minimize(cross_entropy)
sess = tf.InteractiveSession()
init = tf.global_variables_initializer()
sess.run(init)
for _ in range(1000):
    batch_xs, batch_ys = mnist.train.next_batch(100)
    sess.run(train_step, {x: batch_xs, y_: batch_ys})

correct_prediction = tf.equal(tf.argmax(y, 1), tf.argmax(y_, 1))
accuracy = tf.reduce_mean(tf.cast(correct_prediction,
                        tf.float32))
print(sess.run(accuracy, {x: mnist.test.images,
                    y_: mnist.test.labels}))
```

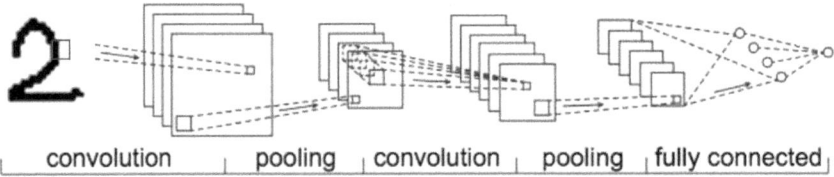

Fig. 4 Convolutional neural network

Constructing Convolutional Neural Network

The image information is extracted continuously through convolution and pooling, and finally the output of the result is realized in the full connection. At the same time, considering the need to have a certain generalization ability, the convolutional neural network will add dropout to prevent overfitting (Fig. 4).

Python code

```python
#Main code of mnist_test.py
def Record_Tensor(tensor, name):
    print("Recording_tensor_" + name + "_...")
    f = open('./my_record/' + name + '.dat', 'w')
    array = tensor.eval()
    if np.size(np.shape(array)) == 1:
        Record_Array1D(array, name, f)
    else:
        if np.size(np.shape(array)) == 2:
            Record_Array2D(array, name, f)
        else:
            if np.size(np.shape(array)) == 3:
                Record_Array3D(array, name, f)
            else:
                Record_Array4D(array, name, f)
    f.close()
def Record_Array1D(array, name, f):
    ...
def Record_Array2D(array, name, f):
    ...
def Record_Array3D(array, name, f):
    ...
def Record_Array4D(array, name, f):
    ...
with tf.name_scope('input'):
```

```
x = tf.compat.v1.placeholder("float", shape=[None, 784])
y_ = tf.compat.v1.placeholder("float", shape=[None, 10])
```

Neural network construction is based on TensorFlow.

Python code

```python
#Main code of mnist_test.py
def weight_variable(shape):
    ...
def bias_variable(shape):
    ...
def conv2d(x, W):
    ...
def max_pool_2x2(x):
    ...
# First Convolutional Layer
with tf.name_scope('1st_CNN'):
    ...
# Second Convolutional Layer
with tf.name_scope('2rd_CNN'):
    ...
# Densely Connected Layer
with tf.name_scope('Densely_NN'):
    ...
# Dropout
with tf.name_scope('Dropout'):
    ...
# Readout Layer
with tf.name_scope('Softmax'):
    ...
with tf.name_scope('Loss'):
    ...
with tf.name_scope('Train'):
    ...
with tf.name_scope('Accuracy'):
    ...
# merged = tf.merge_all_summaries()
merged = tf.compat.v1.summary.merge_all()
writer = tf.compat.v1.summary.FileWriter("logs/", sess.graph)
tf.compat.v1.global_variables_initializer().run()
    ...
```

print("test␣accuracy␣%g" % accuracy.eval(feed_dict={x:
mnist.test.images, y_: mnist.test.labels, rate: 1.0}))

File Format Conversion

The .dat file stored in neural network cannot be read directly by C, and the filename
needs to be converted, in which case the C code shown below can be used.

C code

```c
#Main code of dat2bin.c

char* filename_to_bin(char *filename_i)
{
    int filename_length=0;
    while(filename_i[filename_length]!='\0')
    filename_length++;
    //printf("filename length=%d\n",filename_length);
    char *filename_bin=(char *)malloc(filename_length+1);
    int i=0;
    while(!(filename_i[i]=='.' && filename_i[i+1]=='d'&&
    filename_i[i+2]=='a' && filename_i[i+3]=='t' &&
    filename_i[i+4]=='\0'))//not '.dat\0'
    {
        if(i==filename_length-1)
        {
            free(filename_bin);
            filename_bin=NULL;
            return filename_bin;
        }
        filename_bin[i]=filename_i[i];
        i++;
    }
    filename_bin[i]='.';filename_bin[i+1]='b';
    filename_bin[i+2]='i';filename_bin[i+3]='n';
    filename_bin[i+4]='\0';
    return filename_bin;
}

int main(int argc, char *argv[])
{
```

```
for(int i=1;i<argc;i++)
{
    char *filename_i=argv[i];
    char *filename_bin=filename_to_bin
    (filename_i);
    if(filename_bin==NULL)
    {
        printf("%s is not a dat file\n",
        filename_i);
        break;
    }

    FILE *fp_IN;
    if((fp_IN=fopen(filename_i,"r"))==NULL)
    {
        printf("File %s cannot be opened/n",
        filename_i);
        break;
    }

    FILE* fp_OUT = fopen(filename_bin,"wb");
    if (fp_OUT == NULL)
    {
        printf("File %s cannot be created/n",
        filename_bin);
        break;
    }

    char str[20];
    while(fgets(str,20,fp_IN))
    {
        //printf("%s=%f\n",str,atof(str));
        float tp=atof(str);
        fwrite(&tp,sizeof(float),1,fp_OUT);
    };

    fclose(fp_IN);
    fflush(fp_OUT);
    fclose(fp_OUT);
    free(filename_bin);
}
return 0;
}
```

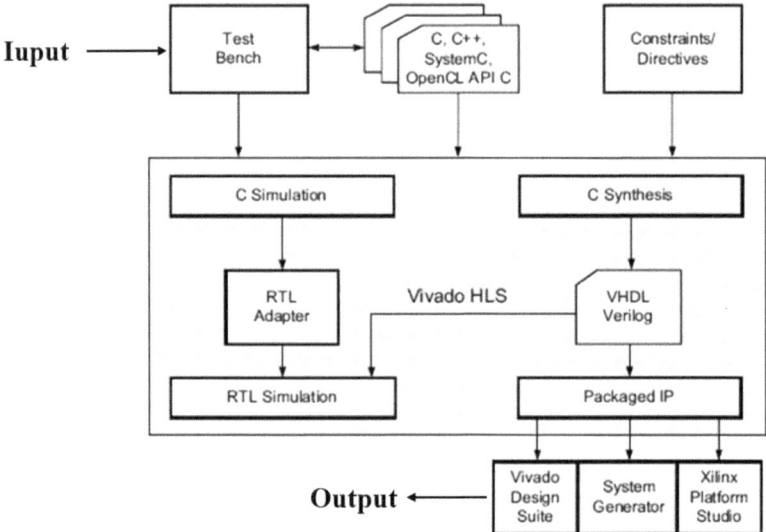

Fig. 5 Vivado HLS

Hardware Section

The above neural network has a large amount of computation and can be executed more quickly through a parallelized design on the hardware side. The following sections are used to write code on the hardware side (Fig. 5).

MNIST Hardware Implementation Ideas

The neural network is mainly Conv and Pool. And then the Conv and Pool circuits that can be used universally in hardware are designed, and CPU and memory are used to control and store data. This will greatly improve the versatility of the design circuit and its parallelism. The design idea is as follows:

Control Conv and Pool through the CPU. First store the picture to be read in the memory, and then the CPU controls Conv to perform the first layer of convolution operations and store the calculation results in the memory. After this step is completed, the CPU reads the calculation result of the last Conv according to the neural network design control Pool and stores the calculated result in the memory again. Thereafter, only the above steps need to be repeated until the operation is completed (Fig. 6).

Implement two subfunctions:

One is responsible for Conv with ReLU.

One is responsible for Pool.

The idea of the C language to implement Conv and Pool parts:

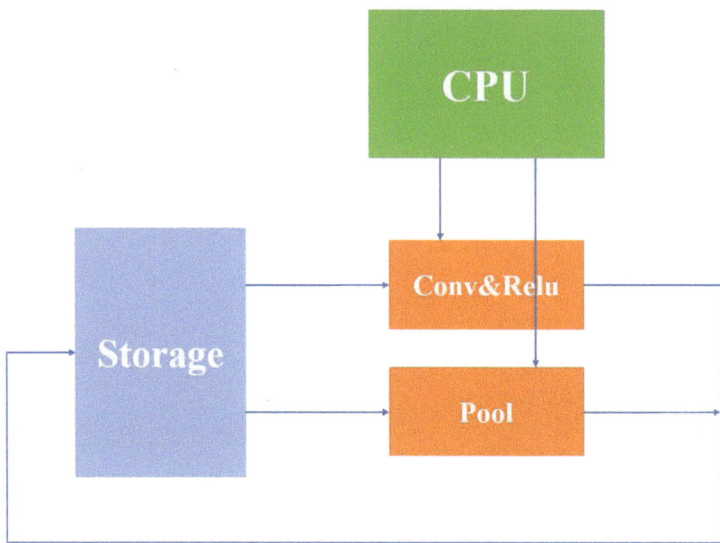

Fig. 6 Hardware implementation framework

For the Conv part, it is necessary to determine the parameters that need to be passed in because it is to design a universal circuit. win and hin represent the width and height of the input layer, respectively, x_stride and y_ stride are used to represent the stride of the convolutional kernel moving in the X and Y directions, while Cin and Cout represent the number of channels entered and the number of channels output, kx and ky are used to represent the size of the convolutional kernel, wout and hout are used to represent the size of the output layer. In addition to the need to specify whether Conv needs to do padding and then the ReLU operation, but also needs the necessary input and output. The Pool part is basically the same as it is (Fig. 7).

Code Writing and Hardware Synthesis

The code for the Conv part is as follows:

Main code of conv_core.cpp

```
#include "conv_core.h"

void Conv(ap_uint<16> CHin,ap_uint<16> Hin,
    ap_uint<16> Win,ap_uint<16> CHout,ap_uint<8> Kx,
    ap_uint<8> Ky,ap_uint<8> Sx,ap_uint<8> Sy,
    ap_uint<1> mode,ap_uint<1> relu_en,
```

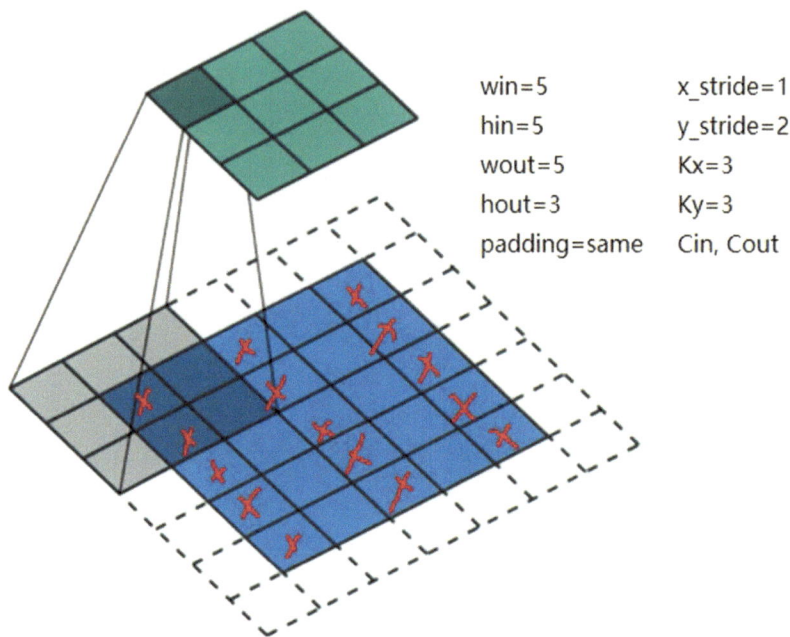

win=5	x_stride=1
hin=5	y_stride=2
wout=5	Kx=3
hout=3	Ky=3
padding=same	Cin, Cout

Fig. 7 C language implementation of conv

```
Dtype_f feature_in[],Dtype_w W[],
Dtype_w bias[],Dtype_f feature_out[]
)//mode: 0:VALID, 1:SAME
{
    ap_uint<8> pad_x,pad_y;
    if(mode==0)
    {
        pad_x=0;pad_y=0;
    }
    else
    {
        pad_x=(Kx−1)/2;pad_y=(Ky−1)/2;
    }
    ap_uint<16> Hout,Wout;
    Wout=(Win+2*pad_x−Kx)/Sx+1;
    Hout=(Hin+2*pad_y−Ky)/Sy+1;

    for(int cout=0;cout<CHout;cout++)
    for(int i=0;i<Hout;i++)
    for(int j=0;j<Wout;j++)
    {
```

```
Dtype_acc sum=0;
for(int ii=0;ii<Ky;ii++)
for(int jj=0;jj<Kx;jj++)
{
ap_int<16> h=i*Sy−pad_y+ii;
ap_int<16> w=j*Sx−pad_x+jj;
if(h>=0 && w>=0 && h<Hin && w<Win)
{
for(int cin=0;cin<CHin;cin++)
{
ii*Kx*CHin*CHout+jj*CHin*CHout+cin*CHout+cout<<"]\n";
Dtype_mul tp=feature_in[h*CHin*Win+w*CHin+cin]*
W[ii*Kx*CHin*CHout+jj*CHin*CHout+cin*CHout+cout];
sum+=tp;
}
}
}
sum+=bias[cout];
if(relu_en & sum<0)
    sum=0;
    feature_out[i*Wout*CHout+j*CHout+cout]=sum;
        }
}
```

The code for the Pool section is as follows:

Main code of pool_core.cpp

```
#include "pool_core.h"

#define max(a,b) ((a>b)?a:b)
#define min(a,b) ((a>b)?b:a)

void Pool(ap_uint<16> CHin,ap_uint<16> Hin,ap_uint<16> Win,
      ap_uint<8> Kx,ap_uint<8> Ky,ap_uint<2> mode,
      Dtype_f feature_in[],Dtype_f feature_out[]
   )//mode: 0:MEAN, 1:MIN, 2:MAX
{
    ap_uint<16> Hout,Wout;
    Wout=Win/Kx;
    Hout=Hin/Ky;
```

```
for(int c=0;c<CHin;c++)
for(int i=0;i<Hout;i++)
for(int j=0;j<Wout;j++)
{
Dtype_f sum;
    if(mode==0)
        sum=0;
        else
if(mode==1)
sum=99999999999999999;
    else
sum=-99999999999999999;
for(int ii=0;ii<Ky;ii++)
for(int jj=0;jj<Kx;jj++)
{
ap_int<16> h=i*Ky+ii;
ap_int<16> w=j*Kx+jj;
switch(mode)
{
case 0:{sum+=feature_in[h*CHin*Win+w*CHin+c];break;}
case 1:{sum=min(sum,feature_in[h*CHin*Win+w*CHin+c]);break;}
case 2:{sum=max(sum,feature_in[h*CHin*Win+w*CHin+c]);break;}
default:break;
}
}
if(mode==0)
sum=sum/(Kx*Ky);
feature_out[i*Wout*CHin+j*CHin+c]=sum;
}
}
```

Add constraints for Conv and Pool Fig. 8.

Construction of Hardware Platforms

Vivado generates an IP core as follows: Click the icon in the red box to generate an IP core (Figs. 9, 10 and 11).

After creating a new project, click on the Create Block Design. Then follow the image below to guide it into the ARM section. Click run block automatic, at which point the DDR interface will be automatically generated.

Click Project Settings, and click IP->Repository Manager to add the previously generated IP core path.

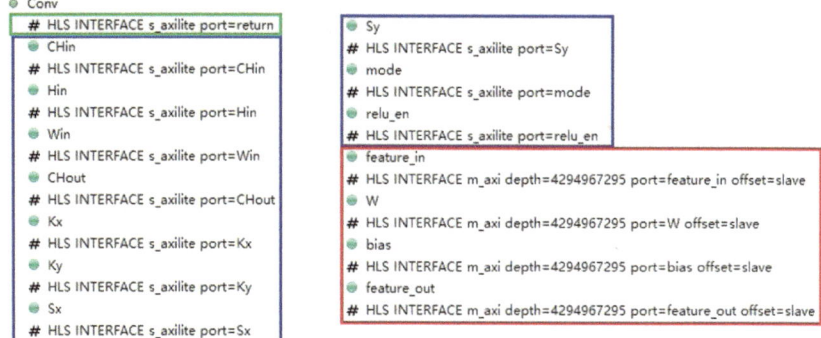

Fig. 8 HLS constraints

Fig. 9 Vivado generates IP core

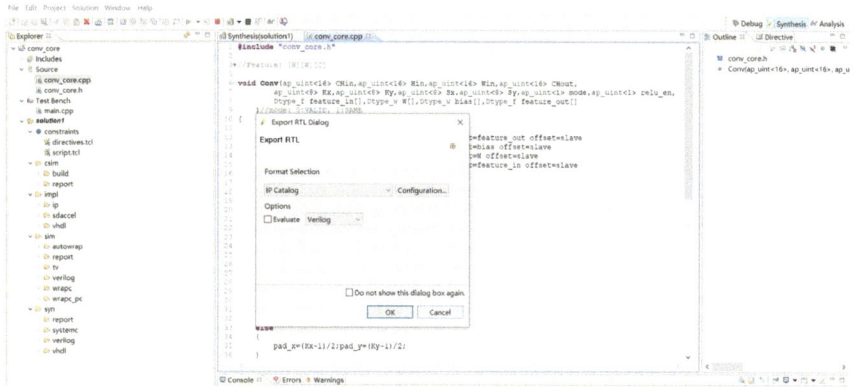

Fig. 10 Vivado generates IP core

Then click run block automatic to connect Conv and Pool slave to ARM.

At this point, the connection between the Conv and Pool control ports has been completed, but it is also necessary to connect the master interface to the ARM. And double-click ARM, the following block diagram will pop up: At this time, you can select one (HP0) or can also choose two (HP0 and HP1) (Fig. 12 and 13).

Note: This step cannot be done automatically through run block automatic like the previous slave port link, because ARM's slave port is not displayed, and you need to manually adjust the display.

As shown in the following figure, verify that the circuit is correct and complete the circuit connection.

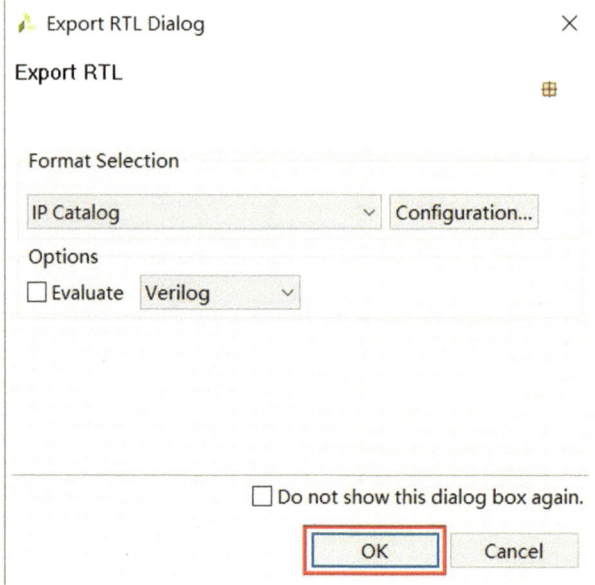

Fig. 11 Vivado generates IP core

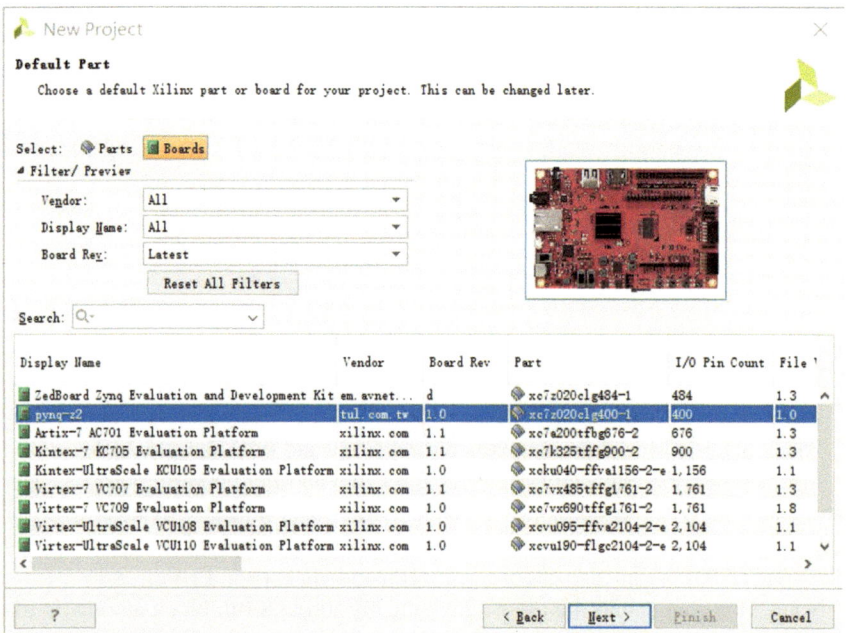

Fig. 12 Import PYNQ-Z2

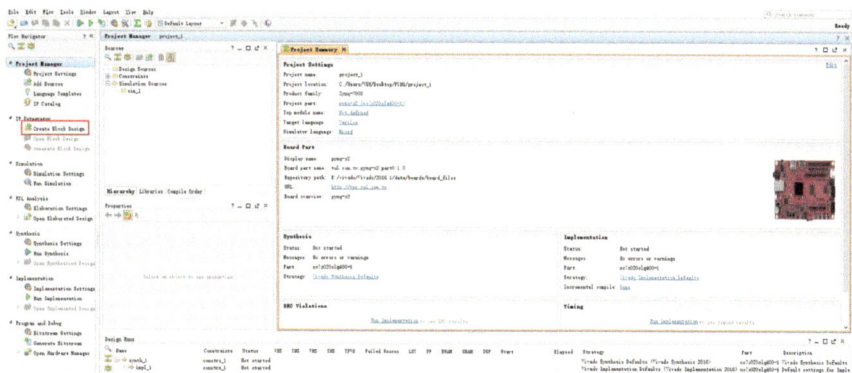

Fig. 13 Create block design

Generate bitstream files that can be transferred to the board. Firstly, click Source->Desing_1->Create HDL Wrapper. Secondly, click Desing_1-> Generate Output Products->. Finally, click Generate Bitstream to generate a bitstream file. And click file->Export-> Export Block Design, generating a .tcl ends with the file, after which you also need to copy both files into the PYNQ board's SD card for transfer (Fig. 14).

Hardware and Software Collaboration

Hardware Driver Writing

For people without PYNQ development experience, you can first try to call the official IP for LED lighting to familiarize, familiar with the basic PYNQ-IP call process can be developed CPU drivers (Figs. 15, 16, 17, 18, 19, 20 and 21).

PYNQ hardware drivers can be written by Python.

Run the driver file conv.py on the PYNQ board.

Python code

#Main code of conv.py
```
import time
from pynq import Overlay
import numpy as np
from pynq import Xlnk
...
xlnk=Xlnk();

ol=Overlay("mnist.bit")
```

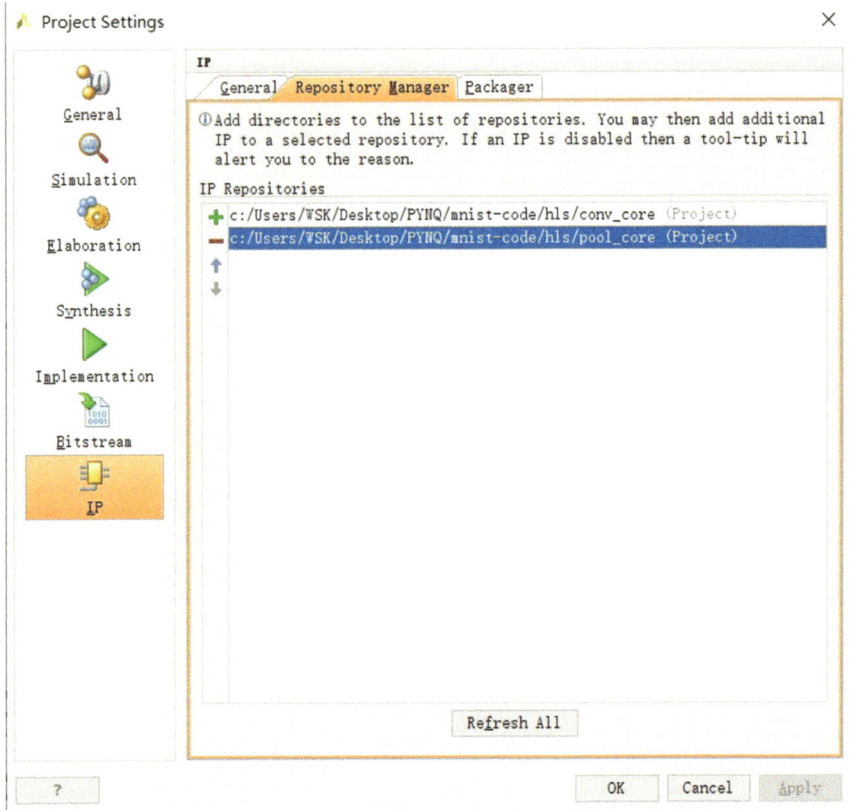

Fig. 14 Add the generated IP core path

ol.ip_dict
ol.download()
conv=ol.Conv_0
print("Overlay␣download␣finish");

feature_in=xlnk.cma_array(shape=(IN_HEIGHT,
IN_WIDTH,IN_CH),cacheable=0,dtype=np.float32)
W=xlnk.cma_array(shape=(KERNEL_HEIGHT,
KERNEL_WIDTH,IN_CH,OUT_CH),cacheable=0,dtype=np.float32)
bias=xlnk.cma_array(shape=(OUT_CH),cacheable=0,
                    dtype=np.float32)
feature_out=xlnk.cma_array(shape=(OUT_HEIGHT,
OUT_WIDTH,OUT_CH),cacheable=0,dtype=np.float32)

#Initialize the feature_in, W, bias

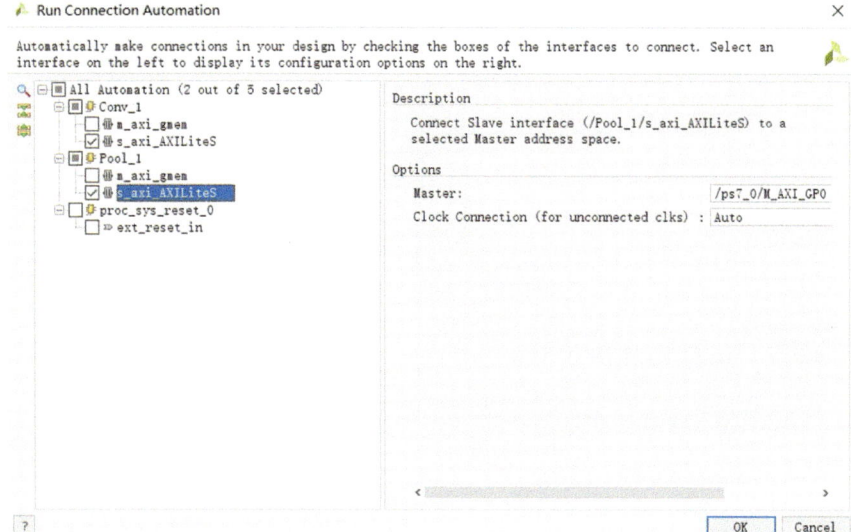

Fig. 15 Run connection automation

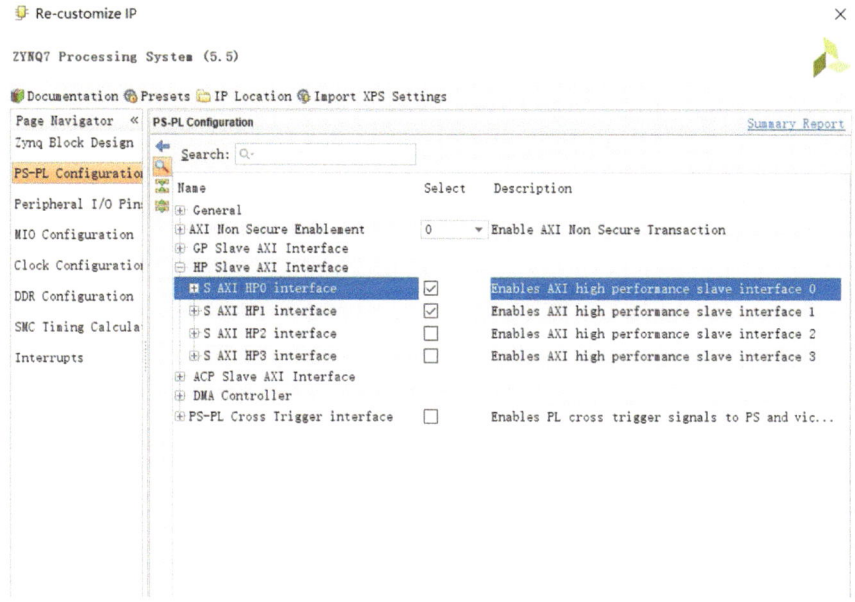

Fig. 16 Choose interface

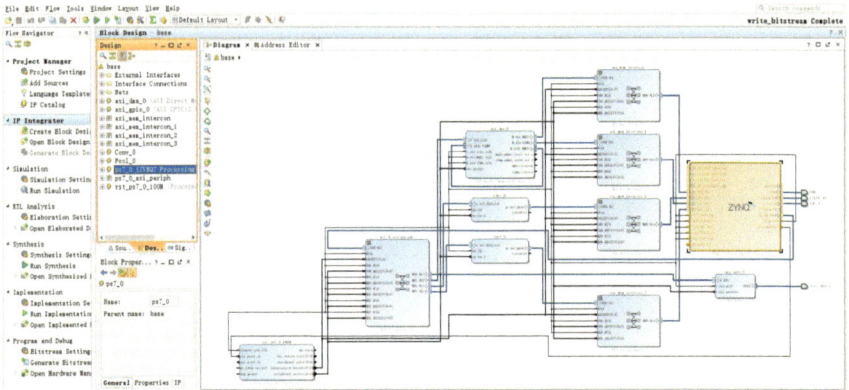

Fig. 17 Complete circuit diagram

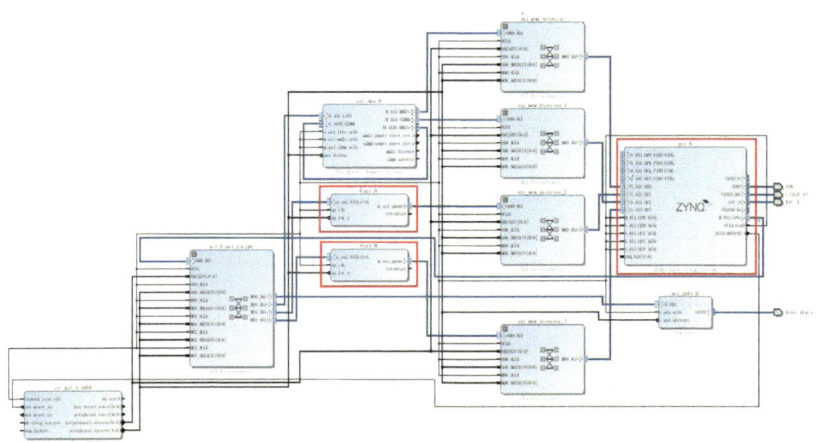

Fig. 18 HComplete circuit diagram

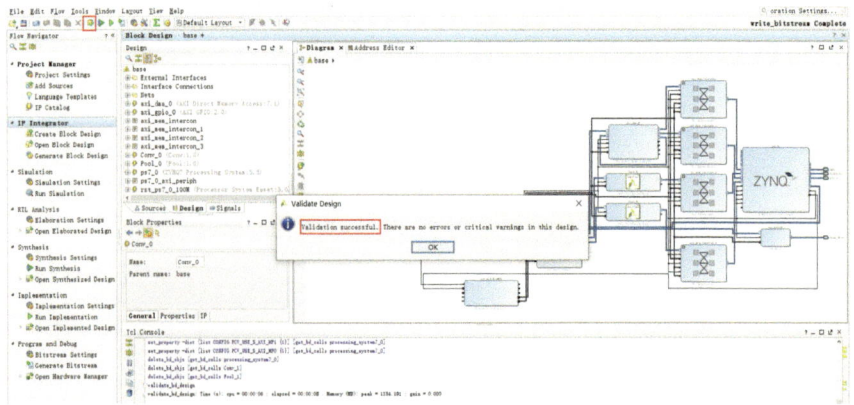

Fig. 19 Validation successful

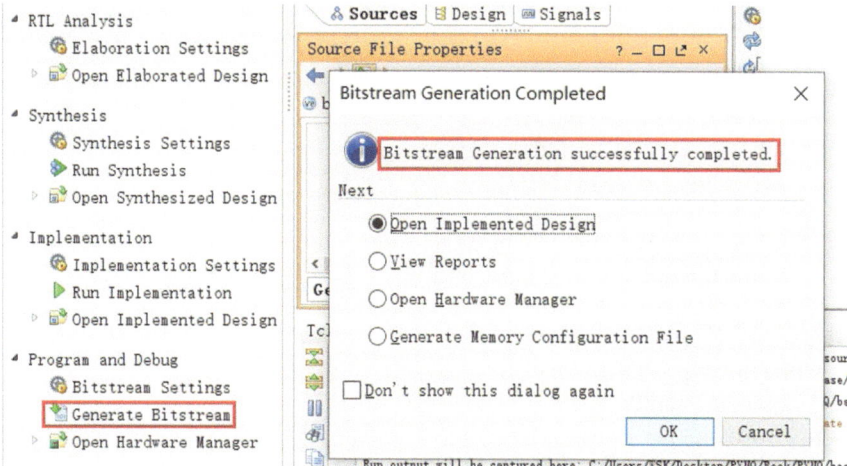

Fig. 20 Generate bitstream

```
root@pynq:/home/xilinx/jupyter_notebooks/mnist# python3 conv.py
/usr/local/lib/python3.6/dist-packages/pynq/overlay.py:299: UserWarning:
  is recommended.
    warnings.warn(message, UserWarning)
Overlay download finish
6
Hardware Run Finish
feature_out[0][0][0]=582.000000
feature_out[0][1][0]=618.000000
feature_out[0][2][0]=654.000000
feature_out[0][3][0]=690.000000
feature_out[0][4][0]=726.000000
feature_out[0][5][0]=762.000000
feature_out[0][6][0]=798.000000
feature_out[0][7][0]=834.000000
feature_out[1][0][0]=942.000000
```

Fig. 21 Execute conv.py

```
for i in range(IN_HEIGHT):
    for j in range(IN_WIDTH):
        for k in range(IN_CH):
            feature_in[i][j][k]=(i*IN_WIDTH+j)*1;

for i in range(KERNEL_HEIGHT):
    for j in range(KERNEL_WIDTH):
        for k in range(IN_CH):
            for l in range(OUT_CH):
                W[i][j][k][l]=(i*KERNEL_WIDTH+j)*1;

for i in range(OUT_CH):
```

```
    bias[i]=i;

for i in range(OUT_HEIGHT):
    for j in range(OUT_WIDTH):
        for k in range(OUT_CH):
            feature_out[i][j][k]=−1;

def RunConv(conv,Kx,Ky,Sx,Sy,mode,relu_en,
            feature_in,W,bias,feature_out):
    ...
    tp=conv.read(0)
    while not ((tp>>1)&0x1):
        tp=conv.read(0);
    print(tp);

RunConv(conv,KERNEL_WIDTH,KERNEL_HEIGHT,X_STRIDE,
Y_STRIDE,MODE,RELU_EN,feature_in,W,bias,feature_out);

print("Hardware_Run_Finish");

for i in range(OUT_HEIGHT):
    for j in range(OUT_WIDTH):
        for k in range(OUT_CH):
            print("feature_out[%d][%d][%d]=%f"%
            (i,j,k,feature_out[i][j][k]));
```

Run the driver file pool.py on the PYNQ board (Figs. 22 and Fig. 23).

Python code

```
#Main code of pool.py
import time
from pynq import Overlay
import numpy as np
from pynq import Xlnk
...
xlnk=Xlnk();

ol=Overlay("mnist.bit")
ol.ip_dict
ol.download()
```

```
pool=ol.Pool_0
print("Overlay.download.finish");

feature_in=xlnk.cma_array(shape=(IN_HEIGHT,IN_WIDTH,IN_CH),
                          cacheable=0,dtype=np.float32)
feature_out=xlnk.cma_array(shape=(OUT_HEIGHT,OUT_WIDTH,OUT_CH),
                           cacheable=0,dtype=np.float32)

#Initialize the feature_in
for i in range(IN_HEIGHT):
    for j in range(IN_WIDTH):
        for k in range(IN_CH):
            feature_in[i][j][k]=(i*IN_WIDTH+j)*1;

for i in range(OUT_HEIGHT):
    for j in range(OUT_WIDTH):
        for k in range(OUT_CH):
            feature_out[i][j][k]=-1;

def RunPool(pool,Kx,Ky,mode,feature_in,feature_out):
    ...

RunPool(pool,KERNEL_WIDTH,KERNEL_HEIGHT,MODE,
                  feature_in,feature_out);

print("Hardware.Run.Finish");

for i in range(OUT_HEIGHT):
    for j in range(OUT_WIDTH):
        for k in range(OUT_CH):
            print("feature_out[%d][%d][%d]=%f"%
            (i,j,k,feature_out[i][j][k]));
```

Demo

The handwritten numeral to be recognized is 2.

Connect the PYNQ board, open terminal in Jupyter, and enter home/Xilinx/jupyter_notebooks/overlay path, performing python3 mnist.py (Fig. 24).

```
root@pynq:/home/xilinx/jupyter_notebooks/mnist# python3 pool.py
/usr/local/lib/python3.6/dist-packages/pynq/overlay.py:299: UserWarning:
  is recommended.
    warnings.warn(message, UserWarning)
Overlay download finish
Hardware Run Finish
feature_out[0][0][0]=0.000000
feature_out[0][1][0]=3.000000
feature_out[0][2][0]=6.000000
feature_out[1][0][0]=30.000000
feature_out[1][1][0]=33.000000
feature_out[1][2][0]=36.000000
feature_out[2][0][0]=60.000000
feature_out[2][1][0]=63.000000
feature_out[2][2][0]=66.000000
```

Fig. 22 Hardware implementation framework

Fig. 23 Handwritten numeral to be recognized

```
root@pynq:/home/xilinx# cd jupyter_notebooks/
root@pynq:/home/xilinx/jupyter_notebooks# cd mnist/
root@pynq:/home/xilinx/jupyter_notebooks/mnist# python3 mnist.py
/usr/local/lib/python3.6/dist-packages/pynq/overlay.py:299: UserWarn
  is recommended.
    warnings.warn(message, UserWarning)
Overlay download finish
Finish initial
input enter to continue
Read image
Finish reading image
Hardware run finish
The number you write is 2
```

Fig. 24 Result of recognition

Python code

```
#Main code of mnist.py
def readbinfile(filename,size):
    ...
def RunConv(conv,Kx,Ky,Sx,Sy,mode,relu_en,
            feature_in,W,bias,feature_out):
    ...
def RunPool(pool,Kx,Ky,mode,feature_in,feature_out):
    ...
#Conv1
IN_WIDTH1=28
IN_HEIGHT1=28
IN_CH1=1

KERNEL_WIDTH1=3
KERNEL_HEIGHT1=3
X_STRIDE1=1
Y_STRIDE1=1

RELU_EN1=1
MODE1=1 #0:VALID, 1:SAME
if(MODE1):
    X_PADDING1=int((KERNEL_WIDTH1−1)/2)
    Y_PADDING1=int((KERNEL_HEIGHT1−1)/2)
else:
    X_PADDING1=0
    Y_PADDING1=0

OUT_CH1=16
OUT_WIDTH1=int((IN_WIDTH1+2*X_PADDING1−KERNEL_WIDTH1)
            /X_STRIDE1+1)
OUT_HEIGHT1=int((IN_HEIGHT1+2*Y_PADDING1−KERNEL_HEIGHT1)
            /Y_STRIDE1+1)


#Pool1
MODE11=2 #mode: 0:MEAN, 1:MIN, 2:MAX
IN_WIDTH11=OUT_WIDTH1
IN_HEIGHT11=OUT_HEIGHT1
IN_CH11=OUT_CH1

KERNEL_WIDTH11=2
KERNEL_HEIGHT11=2

OUT_CH11=IN_CH11
OUT_WIDTH11=int(IN_WIDTH11/KERNEL_WIDTH11)
OUT_HEIGHT11=int(IN_HEIGHT11/KERNEL_HEIGHT11)

#Conv2
...
#Pool2
```

```
...

#Fc1
IN_WIDTH3=OUT_WIDTH21
IN_HEIGHT3=OUT_HEIGHT21
IN_CH3=OUT_CH21

KERNEL_WIDTH3=7
KERNEL_HEIGHT3=7
X_STRIDE3=1
Y_STRIDE3=1

RELU_EN3=1
MODE3=0 #0:VALID, 1:SAME
if(MODE3):
    X_PADDING3=int((KERNEL_WIDTH3−1/2))
    Y_PADDING3=int((KERNEL_HEIGHT3−1)/2)
else:
    X_PADDING3=0
    Y_PADDING3=0

OUT_CH3=128
OUT_WIDTH3=int((IN_WIDTH3+2*X_PADDING3−KERNEL_WIDTH3)
        /X_STRIDE3+1)
OUT_HEIGHT3=int((IN_HEIGHT3+2*Y_PADDING3−KERNEL_HEIGHT3)
        /Y_STRIDE3+1)

#Fc2
...

xlnk=Xlnk();

ol=Overlay("mnist.bit")
ol.ip_dict
ol.download()
conv=ol.Conv_0
pool=ol.Pool_0
print("Overlay_download_finish");

#input image
image=xlnk.cma_array(shape=(IN_HEIGHT1,IN_WIDTH1,IN_CH1),
            cacheable=0,dtype=np.float32)

#conv1
W_conv1=xlnk.cma_array(shape=(KERNEL_HEIGHT1,KERNEL_WIDTH1,
        IN_CH1,OUT_CH1),cacheable=0,dtype=np.float32)
b_conv1=xlnk.cma_array(shape=(OUT_CH1),cacheable=0,
                dtype=np.float32)
h_conv1=xlnk.cma_array(shape=(OUT_HEIGHT1,OUT_WIDTH1,OUT_CH1),
        cacheable=0,dtype=np.float32)
h_pool1=xlnk.cma_array(shape=(OUT_HEIGHT11,OUT_WIDTH11,OUT_CH11),
        cacheable=0,dtype=np.float32)
```

```
#conv2
...

#fc1
W_fc1=xlnk.cma_array(shape=(KERNEL_HEIGHT3, KERNEL_WIDTH3,
IN_CH3, OUT_CH3),cacheable=0,dtype=np.float32)
b_fc1=xlnk.cma_array(shape=(OUT_CH3),cacheable=0,dtype=np.float32)
h_fc1=xlnk.cma_array(shape=(OUT_HEIGHT3,OUT_WIDTH3,OUT_CH3),
        cacheable=0,dtype=np.float32)

#fc2
...

#Initialize W, bias

w_conv1=readbinfile("/home/xilinx/overlay/mnist
/data/W_conv1.bin",KERNEL_HEIGHT1*KERNEL_WIDTH1*IN_CH1*OUT_CH1)
w_conv1=w_conv1.reshape((KERNEL_HEIGHT1,KERNEL_WIDTH1,
                IN_CH1,OUT_CH1))
...
B_conv1=readbinfile("/home/xilinx/overlay/mnist
/data/b_conv1.bin",OUT_CH1)
...

#w_conv2
...
#B_conv2
...

w_fc1=readbinfile("/home/xilinx/overlay/mnist
/data/W_fc1.bin",KERNEL_HEIGHT3*KERNEL_WIDTH3*IN_CH3*OUT_CH3)
w_fc1=w_fc1.reshape((KERNEL_HEIGHT3,KERNEL_WIDTH3,IN_CH3,OUT_CH3))
for i in range(KERNEL_HEIGHT3):
    for j in range(KERNEL_WIDTH3):
        for k in range(IN_CH3):
            for l in range(OUT_CH3):
                W_fc1[i][j][k][l]=w_fc1[i][j][k][l]
B_fc1=readbinfile("/home/xilinx/overlay/mnist/data/b_fc1.bin",OUT_CH3)
...

#w_fc2
...
#B_fc2
...

print("Finish_initial")

while(1):
    while(1):
        g=input("input_enter_to_continue")
        break
    image1=cv2.imread("/home/xilinx/overlay/mnist
_____/data/1.jpg",cv2.IMREAD_GRAYSCALE).astype(np.float32)
```

```
print("Read_image")
#image1=image1.reshape((IN_HEIGHT1,IN_WIDTH1,IN_CH1))
for i in range(IN_HEIGHT1):
    for j in range(IN_WIDTH1):
        for k in range(IN_CH1):
            image[i][j][k]=(255-image1[i][j])/255
print("Finish_reading_image")
#conv1
RunConv(conv,KERNEL_WIDTH1,KERNEL_HEIGHT1,X_STRIDE1,
Y_STRIDE1,MODE1,RELU_EN1,image,W_conv1,b_conv1,h_conv1)
RunPool(pool, KERNEL_WIDTH11, KERNEL_HEIGHT11,
MODE11, h_conv1, h_pool1)
# conv2
...
# fc1
RunConv(conv, KERNEL_WIDTH3, KERNEL_HEIGHT3, X_STRIDE3,
Y_STRIDE3, MODE3, RELU_EN3, h_pool2, W_fc1, b_fc1,
     h_fc1)
# fc2
...

print("Hardware_run_finish")
MAX = h_fc2[0][0][0]
result=0
for i in range(1,OUT_CH4):
    if(h_fc2[0][0][i]>MAX):
        MAX=h_fc2[0][0][i]
        result=i
print("The_number_you_write_is_"+str(result))
```

We write Python code according to the LeNet-5 model and call the previously conv.py and pool.py part of the code, loading mnist.bit and mnist.tcl file, to build the final functional circuit, and import the training parameters of the network built by TensorFlow before.

? Questions

Now understand the theory of hardware accelerated neural network?

Video Super Resolution Based on PYNQ

Video super resolution (VSR) system can improve video quality. On pure software section, we used NumPy to build a convolutional neural network to improve PSNR.

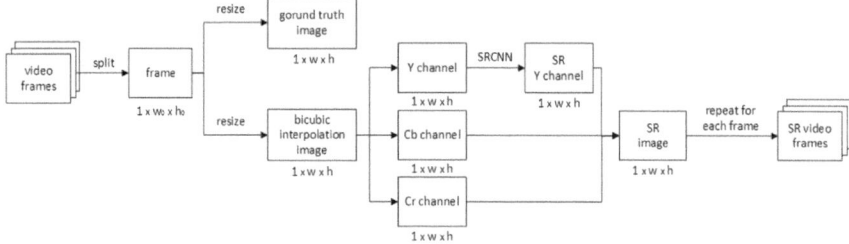

Fig. 25 System flowchart

As shown in Fig. 25, we split video into frames and use super resolution convolutional neural network (SRCNN)[1] to only Y channel.

Experimental development environment

Python 3.7+; numpy; opencv-python;

Software Section

Pre-processing

This section gives examples about how we split video into frames and how to convert images from RGB to YUV.

Python code

```python
# video_splt.py
import cv2
from os import listdir


def get_file_list(file_path: str):
    """
    :param file_path: the file path where you want to get file
    :return: list, files sorted by name
    """
    dir_list = listdir(file_path)
    if not dir_list:
        return
    else:
        # use lambda expression
        # sort the files in ascending order of last modification time
        # os.path.getmtime() get last modified time
```

```python
    # os.path.getctime() get file created time
    # dir_list = sorted(dir_list, key=lambda x: os.path.getmtime(os.path.join(file_path, x)))
    # print(dir_list)
    try:
        # bicubic dir add in SR dir, not in sort range
        dir_list.remove("bicubic")
    except ValueError:
        pass
    dir_list = sorted(dir_list, key=lambda x: int(x[:-4])) # sorted by name
    # print(dir_list)
    return dir_list


def video_split(video_path: str, save_path: str):
    """
    @param video_path: video path
    @param save_path: dir to save split frames
    """
    vc = cv2.VideoCapture(video_path)
    # write_c = 0 # control num of test frames
    c = 0
    if vc.isOpened():
        rval, frame = vc.read()
    else:
        rval = False
    while rval:
        rval, frame = vc.read() # frame: <class 'numpy.ndarray'>
        # extract 5 frames per second
        if c % 5 == 0:
            try:
                cv2.imwrite(save_path + "/" + str('%06d' % c) + '.jpg', frame)
                cv2.waitKey(1)
                # write_c += 1
                # if write_c > 2:
                # break
            except: # current frames is empty
                pass
        c = c + 1
```

Python code

```python
# util.py
import numpy as np
# Source for formulas: \url{https://sistenix.com/rgb2ycbcr.html}


def rgb2y(i):
    if type(i) == np.ndarray:
        return 16. + (65.738 * i[:, :, 0] + 129.057 * i[:, :, 1] + 25.064 * i[:, :, 2]) / 256.
    else:
        raise Exception('Unknown_Type', type(i))


def rgb2ycbcr(i):
    if type(i) == np.ndarray:
        y = 16. + (65.738 * i[:, :, 0] + 129.057 *
```

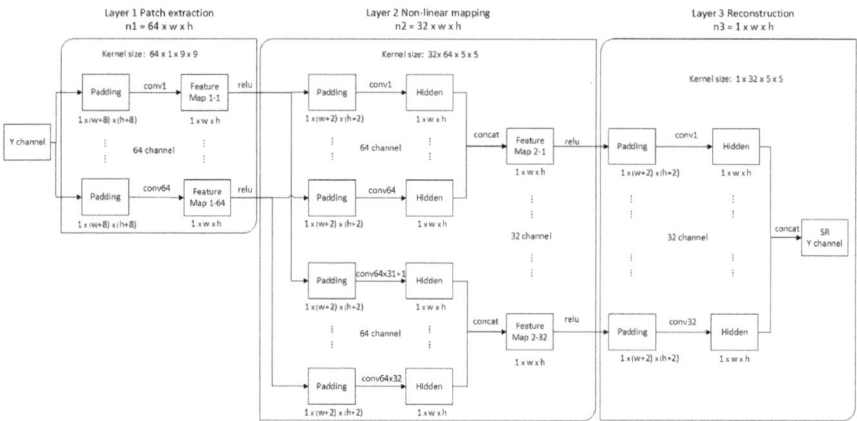

Fig. 26 Network to SR Y channel

$$i[:, :, 1] + 25.064 * i[:, :, 2]) / 256.$$
$$cb = 128. + (-37.945 * i[:, :, 0] - 74.494 *$$
$$i[:, :, 1] + 112.439 * i[:, :, 2]) / 256.$$
$$cr = 128. + (112.439 * i[:, :, 0] - 94.154 *$$
$$i[:, :, 1] - 18.285 * i[:, :, 2]) / 256.$$

```
        return np.array([y, cb, cr])
    else:
        raise Exception('Unknown_Type', type(i))


def ycbcr2rgb(i):
    if type(i) == np.ndarray:
        r = 298.082 * i[0] / 256. + \
            408.583 * i[2] / 256. - 222.921
        g = 298.082 * i[0] / 256. - 100.291 * i[1] \
            / 256. - 208.120 * i[2] / 256. + 135.576
        b = 298.082 * i[0] / 256. + \
            516.412 * i[1] / 256. - 276.836
        return np.array([r, g, b])
    else:
        raise Exception('Unknown_Type', type(i))
```

SR a Single Image

We use NumPy to build SRCNN[1]. Here is the overall network architecture as shown in Fig. 26.

The program below SR is a single image. Here we use the trained mat file from http://mmlab.ie.cuhk.edu.hk/projects/SRCNN/SRCNN_v1.zip. Moreover, we choose model\9-5-5(ImageNet)\x3.mat to test.

Python code

```python
# matlab_SR_pic.py
from PIL import Image
from utility import *
import numpy as np
import scipy.io as io
from scipy import ndimage
from sewar.full_ref import psnr
import os
import time
import prettytable


# In this file we use this fun to get dataset
def get_file_list(file_path: str):
    """
    :param file_path: the file path where you want to get file
    :return: list, files sorted by name
    """
    dir_list = os.listdir(file_path)
    if not dir_list:
        return
    else:
        try:
            dir_list.remove("bicubic") # no need
        except ValueError:
            pass
        return dir_list


def SRCNN_pic(pic: str, output_dir: str, SRCNN_PSNR: list, BICUBIC_PSNR: list, loading_bar: bool):
    """
    :param pic: input low resolution frame
    :param output_dir: directory to save the high resolution frame
    :param SRCNN_PSNR: list to store SRCNN PSNR of each frame
    :param BICUBIC_PSNR: list to store BICUBIC PSNR of each frame
    :param loading_bar: print loading bar or not
    :return: none
    """
    scale = 3 # global scale. There are only weights for scale x3
    image = Image.open(pic).convert('RGB')

    # --- image resizing & preparation ---
    # get a number that is divisible by scale and closest to the actual image size
    image_width = (image.width // scale) * scale
    # same as height
    image_height = (image.height // scale) * scale
    # resize to a multiple of 3 to get the ground truth
    image = image.resize((image_width, image_height), resample=Image.BICUBIC)
    # ground truth, a resized actual image used to calculate PSNR with Bicubic and SRCNN
    ground_truth = np.array(image).astype(np.uint8)
    # actual resizing to 1/3
    # image is the actual input image
    image = image.resize((image.width // scale, image.height // scale), resample=Image.BICUBIC)
    # back x3
    image = image.resize((image.width * scale, image.height * scale), resample=Image.BICUBIC)
    pic_name = pic.split("/")[-1] # "000000.jpg"
    # use bicubic to SR the input image
    image.save(output_dir + "bicubic/" + pic_name)
    if os.path.exists(output_dir + "bicubic/"): # dir exists
```

```
            pass
        else:
            try:
                os.makedirs(output_dir + "bicubic/")
            except FileExistsError: # dir exists
                pass

    image = np.array(image).astype(np.uint8)
    ycbcr = rgb2ycbcr(image)

    y = ycbcr[0]
    y /= 255. # scaling pixel values from 0-255 to 0-1

    # load CNN model parameters
    matr = io.loadmat(r'x3.mat')
    # print(matr.keys())
    weights_conv1 = matr['weights_conv1'] # (81,64)
    biases_conv1 = matr['biases_conv1'] # (64,1)
    weights_conv2 = matr['weights_conv2'] # (64,25,32)
    biases_conv2 = matr['biases_conv2'] # (32,1)
    weights_conv3 = matr['weights_conv3'] # (32,25)
    biases_conv3 = matr['biases_conv3'] # (1,1)
    [conv1_patchsize2, conv1_filters] = np.shape(weights_conv1)
    conv1_patchsize = int(np.sqrt(conv1_patchsize2))
    [conv2_channels, conv2_patchsize2, conv2_filters] = np.shape(weights_conv2)
    conv2_patchsize = int(np.sqrt(conv2_patchsize2))
    [conv3_channels, conv3_patchsize2] = np.shape(weights_conv3)
    conv3_patchsize = int(np.sqrt(conv3_patchsize2))
    [hei, wid] = np.shape(y)

    sum = conv1_filters + conv2_filters * conv2_channels + conv3_channels
    calculator = 0
    # conv layer 1
    weights_conv1 = np.reshape(weights_conv1, (conv1_patchsize, conv1_patchsize, conv1_filters))
    conv1_data = np.zeros((hei, wid, conv1_filters))
    for i in range(conv1_filters):
        # image = np.pad(image, padding, 'edge') # copy the edge of the image to padding
        conv1_data[:, :, i] = ndimage.convolve(y, weights_conv1[:, :, i])
        # relu
        conv1_data[:, :, i] += np.full((hei, wid), biases_conv1[i, 0])
        conv1_data[:, :, i] = (abs(conv1_data[:, :, i]) + conv1_data[:, :, i]) / 2
        if loading_bar:
            calculator += 1
            part = 100 * (calculator) / sum
            have_done = '*' * int(part)
            to_do = '.' * (100 - int(part))
            print('\r{}{}Loading:_{:.2f}%'.format(have_done, to_do, part), end='')

    # conv layer 2
    conv2_data = np.zeros((hei, wid, conv2_filters))
    for i in range(conv2_filters):
        for j in range(conv2_channels):
            conv2_subfilter = np.reshape(weights_conv2[j, :, i], (conv2_patchsize, conv2_patchsize))
            conv2_data[:, :, i] = conv2_data[:, :, i] + ndimage.convolve(conv1_data[:, :, j], conv2_subfilter)
            if loading_bar:
                calculator += 1
                part = 100 * (calculator) / sum
                have_done = '*' * int(part)
                to_do = '.' * (100 - int(part))
                print('\r{}{}Loading:_{:.2f}%'.format(have_done, to_do, part), end='')
        # relu
        conv2_data[:, :, i] += np.full((hei, wid), biases_conv2[i, 0])
        conv2_data[:, :, i] = (abs(conv2_data[:, :, i]) + conv2_data[:, :, i]) / 2
```

```python
# conv layer 3
conv3_data = np.zeros((hei, wid))
for i in range(conv3_channels):
    conv3_subfilter = np.reshape(weights_conv3[i, :], (conv3_patchsize, conv3_patchsize))
    conv3_data[:, :] = conv3_data[:, :] + ndimage.convolve(conv2_data[:, :, i], conv3_subfilter)
    if loading_bar:
        calculator += 1
        part = 100 * (calculator) / sum
        have_done = '*' * int(part)
        to_do = '.' * (100 − int(part))
        print('\r{}{}Loading:_{:.2f}%'.format(have_done, to_do, part), end='')

# SRCNN reconstruction
im_h = conv3_data[:, :] + biases_conv3

new_pic_arry = np.array([np.squeeze(im_h) * 255.0, ycbcr[1], ycbcr[2]])
new_pic_arry = ycbcr2rgb(new_pic_arry)
output_np = np.clip(new_pic_arry, 0.0, 255.0).astype(np.uint8).transpose([1, 2, 0])
output = Image.fromarray(output_np, 'RGB')
output.save(output_dir + pic_name)

# image metrics
PSNR_srcnn = round(psnr(ground_truth, output_np, 255), 2)
PSNR_bicubic = round(psnr(ground_truth, image, 255), 2)
SRCNN_PSNR.append(PSNR_srcnn)
BICUBIC_PSNR.append(PSNR_bicubic)
if loading_bar:
    print()
    print("SRCNN−PSNR:_%.2f_dB" % PSNR_srcnn, end="__")
    print("BICUBIC−PSNR:_%.2f_dB" % PSNR_bicubic)


if __name__ == '__main__':
    all_start_time = time.perf_counter()
    Set5 = get_file_list("SRCNN−Test/Set5")
    Set14 = get_file_list("SRCNN−Test/Set14")
    Set5_PSNR_srcnn = []
    Set5_PSNR_bicubic = []
    Set5_time = []
    Set14_PSNR_srcnn = []
    Set14_PSNR_bicubic = []
    Set14_time = []
    # print("Name\t", "SRCNN−PSNR(dB)\t", "BICUBIC−PSNR(dB)\t", "Time(s)\t")
    for i in range(len(Set5)):
        start_time = time.perf_counter()
        SRCNN_pic("SRCNN−Test/Set5/" + Set5[i], "SR/", Set5_PSNR_srcnn, Set5_PSNR_bicubic, False)
        end_time = round(time.perf_counter() − start_time, 2)
        Set5_time.append(end_time)

    print("Set5")
    table5 = prettytable.PrettyTable()
    table5.add_column('Name', Set5)
    table5.add_column('SRCNN−PSNR(dB)', Set5_PSNR_srcnn)
    table5.add_column('BUCUBIC−PSNR(dB)', Set5_PSNR_bicubic)
    table5.add_column('Time(s)', Set5_time)
    print(table5)

    for i in range(len(Set14)):
        start_time = time.perf_counter()
        SRCNN_pic("SRCNN−Test/Set14/" + Set14[i], "SR/", Set14_PSNR_srcnn, Set14_PSNR_bicubic, False)
        end_time = round(time.perf_counter() − start_time, 2)
        Set14_time.append(end_time)
```

```
print("Set14")
table14 = prettytable.PrettyTable()
table14.add_column('Name', Set14)
table14.add_column('SRCNN−PSNR(dB)', Set14_PSNR_srcnn)
table14.add_column('BUCUBIC−PSNR(dB)', Set14_PSNR_bicubic)
table14.add_column('Time(s)', Set14_time)
print(table14)
all_end_time = time.perf_counter()
print("Total used time {} secs.".format(all_end_time − all_start_time))
```

Main Function

Finally, we run main.py to SR a video.

Python code

```
# main.py import os
from video_split import *
from matlab_SR_pic import *
import cv2
import time
import numpy as np

start_time = time.perf_counter()

VID = "test5.mp4" # video to be SR
Input_pic_path = "Input_VID_pic/" + VID[0:−4] + "/" # dir to store video frame
Output_pic_path = "Output_VID_pic/" + VID[0:−4] + "/" # dir to store SR video frame
if os.path.exists(Input_pic_path) and get_file_list(Input_pic_path): # already split
    pass
else:
    try:
        os.makedirs(Input_pic_path)
    except FileExistsError: # dir exists
        pass
    print("split video ", VID, "to dir ", Input_pic_path)
    video_split(VID, Input_pic_path) # split video into frames

if not os.path.exists(Output_pic_path):
    os.makedirs(Output_pic_path)
    os.makedirs(Output_pic_path + "bicubic")

# each frames of video, sorted by name
pic_list = get_file_list(Input_pic_path)
pic_count = 0
SRCNN_PSNR = []
BICUBIC_PSNR = []
for pic in pic_list:
    print("{} begins ".format(pic))
    pic_count += 1
    # SR each frame, save to new dir, display loading bar
    SRCNN_pic(Input_pic_path + pic, Output_pic_path, SRCNN_PSNR, BICUBIC_PSNR, True)
    # print()
    # break # for testing single pic SR
```

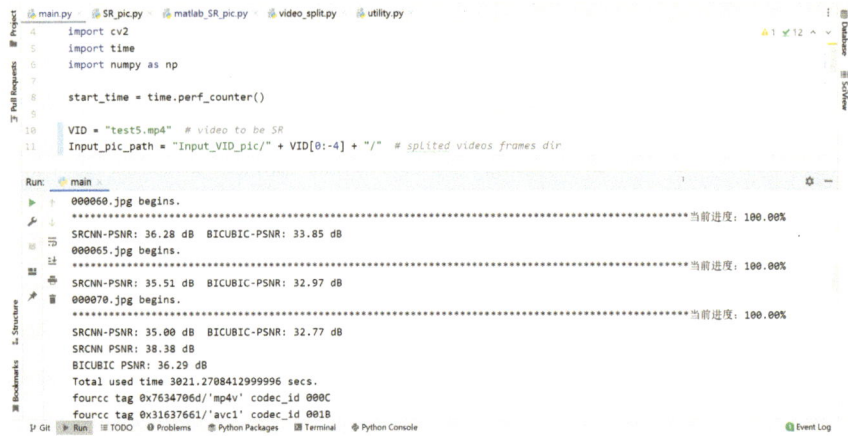

Fig. 27 SR videos console output

```
Set5
+------------------+----------------+------------------+----------+
|      Name        | SRCNN-PSNR(dB) | BUCUBIC-PSNR(dB) | Time(s)  |
+------------------+----------------+------------------+----------+
|   baby_GT.bmp    |     34.55      |      33.16       |  64.48   |
|   bird_GT.bmp    |     32.59      |      30.61       |  11.22   |
| butterfly_GT.bmp |     27.02      |      23.21       |  10.65   |
|   head_GT.bmp    |     32.2       |      31.48       |  10.2    |
|   woman_GT.bmp   |     30.08      |      27.36       |  11.19   |
+------------------+----------------+------------------+----------+

Set14
+------------------+----------------+------------------+----------+
|      Name        | SRCNN-PSNR(dB) | BUCUBIC-PSNR(dB) | Time(s)  |
+------------------+----------------+------------------+----------+
|   baboon.bmp     |     21.85      |      21.47       |  56.45   |
|   barbara.bmp    |     24.97      |      24.67       |  93.56   |
|   bridge.bmp     |     26.1       |      25.12       |  64.29   |
| coastguard.bmp   |     26.03      |      25.26       |  13.6    |
|   comic.bmp      |     23.93      |      22.42       |  12.57   |
|   face.bmp       |     30.51      |      29.93       |  10.96   |
|   flowers.bmp    |     27.73      |      26.12       |  33.3    |
|   foreman.bmp    |     29.53      |      28.12       |  13.57   |
|   lenna.bmp      |     32.41      |      30.83       |  64.17   |
|   man.bmp        |     28.34      |      26.7        |  64.34   |
|   monarch.bmp    |     31.79      |      28.43       | 107.07   |
|   pepper.bmp     |     30.18      |      29.18       |  62.85   |
|   ppt3.bmp       |     26.0       |      23.0        |  75.34   |
|   zebra.bmp      |     28.4       |      25.87       |  51.86   |
+------------------+----------------+------------------+----------+

Total used time 831.6630042659999 secs.

Process finished with exit code 0
```

Fig. 28 SR single picture console output

```
SR_pic_list = get_file_list(Output_pic_path) # SR pic
img = cv2.imread(Output_pic_path + SR_pic_list[0])
size = (img.shape[1], img.shape[0])
# VideoWriter is a video saving method from CV2, to save the synthesized video to this path
# VideoWriter(video_name, encoder, frame_rate, pic_size)
# pic_size should be the same as each frame
video = cv2.VideoWriter('SRCNN−output.mp4', −1, 5, size)
for SR_pic in SR_pic_list:
    img = cv2.imread(Output_pic_path + SR_pic)
    video.write(img)
video.release()
print("SRCNN_PSNR:_%.2f_dB" % np.mean(SRCNN_PSNR))

video = cv2.VideoWriter('BICUBIC−output.mp4', −1, 5, size)
for SR_pic in SR_pic_list:
    img = cv2.imread(Output_pic_path + "bicubic/" + SR_pic)
    video.write(img)
video.release()
print("BICUBIC_PSNR:_%.2f_dB" % np.mean(BICUBIC_PSNR))

end_time = time.perf_counter()
print("Total_used_time_{}_secs.".format(end_time−start_time))
```

Running Results

Run main.py, two directories will be made to store each frames of input videos and SR frames (Figs. 27 and 28).

You can also test single test on single picture instead of videos, run mat-lab_SR_pic.py to test on Set5 and Set14.

Reference

1. C. Dong, C.C. Loy, K. He, X. Tang (2016) Image super-resolution using deep convolutional networks